Phenolic, Sulfur, and Nitrogen Compounds in Food Flavors

George Charalambous, EDITOR
Anheuser-Busch, Inc.

Ira Katz, EDITOR
International Flavors and Fragrances

A symposium sponsored

by the Division of

Agricultural and Food

Chemistry at the 170th

Meeting of the American

Chemical Society,

Chicago, Ill.,

August 25–26, 1975

A C S   S Y M P O S I U M   S E R I E S **26**

AMERICAN CHEMICAL SOCIETY
WASHINGTON, D. C.   1976

Library of Congress CIP Data

Phenolic, sulfur, and nitrogen compounds in food flavors.
    (ACS symposium series; 26 ISSN 0097-6156)

    Includes bibliographical references and index.

    1. Flavoring essences—Congresses.   2. Phenols—Con-
gresses. 3. Flavonoids—Congresses.
    I. Charalambous, George, 1922-   . II. Katz, Ira,
1933-   . III. American Chemical Society. Division of
Agricultural and Food Chemistry.   IV. Series: American
Chemical Society. ACS symposium series; 26.

TP418.P46   664'.06   76-16544
ISBN 0-8412-0330-X   ACSMC8 26 1–215

ACS Symposium Series

Robert F. Gould, *Editor*

FOREWORD

The ACS SYMPOSIUM SERIES was founded in 1974 to provide a medium for publishing symposia quickly in book form. The format of the SERIES parallels that of the continuing ADVANCES IN CHEMISTRY SERIES except that in order to save time the papers are not typeset but are reproduced as they are submitted by the authors in camera-ready form. As a further means of saving time, the papers are not edited or reviewed except by the symposium chairman, who becomes editor of the book. Papers published in the ACS SYMPOSIUM SERIES are original contributions not published elsewhere in whole or major part and include reports of research as well as reviews since symposia may embrace both types of presentation.

CONTENTS

PREFACE

During the past decade there has been a growing interest in all aspects of our food supply. This concern has been precipitated by a variety of factors such as predicted world food shortages, interest in the development of new foods and food analogs, diet and health, consumer concern, etc. In order to solve these problems we must know more about the chemistry of the foods we consume. Implicit in this is a better understanding of flavor chemistry.

In this respect, this volume is very timely since sulfur, nitrogen, and phenolic compounds are important contributors to food flavors. Only recently have we begun to realize how ubiquitous these chemicals are in food systems. The papers in this book will give some insight into the complexity and breadth of this area of food chemistry. Hopefully, it will stimulate the flavor chemist to continue and expand his endeavors.

Anheuser Busch, Inc.　　　　　　　　　　　　GEORGE CHARALAMBOUS
St. Louis, Mo.

International Flavors and Fragrances　　　　　　　　IRA KATZ
Union Beach, N.J.
January 23, 1976

Role of Flavones and Related Compounds in Retarding Lipid–Oxidative Flavor Changes in Foods

DAN E. PRATT

Department of Foods and Nutrition, Purdue University, West Lafayette, Ind. 47907

The term flavonoid is generally used to denote the group of plant phenols characterized by the carbon skeleton C_6-C_3-C_6. The basic structure of these compounds consists of two aromatic rings linked by a three carbon aliphatic chain which normally has been condensed to form a pyran or less commonly a furan ring. As the name implies flavone may be considered the general type compound of the flavonoid group. Based chiefly on the oxidation state of the aliphatic fragment, these compounds may be subdivided into several groups (1, 2, 3). The widest and most inclusive classification (2) places the flavonoids into three classes:

1) The anthoxanthins include all flavonoids that possess a carbonyl group in the 4-position. The center condensed ring may be either the pyran or furan structure; or in one case (the chalcones) the alphatic fragment is not condensed into a ring.
2) The flavans include flavonoids that do not possess a carbonyl in the 4-position. The center condensed ring is always intact and is the pyran structure.
3) The anthocyanins are flavylium salts. These may be considered as flavans in the highest state of oxidation.

These three classes may be further divided as shown in Figures 1 and 2.

Several phenolic compounds that are not flavonoids, but are closely related to flavonoids biosynthetically and in their distribution, must also be considered. These compounds are in the cinnamic acids (3-phenyl propenoic acid derivatives), esters of cinnamic acids and hydroxy and/or methoxy derivative of coumarin.

In the plant kingdom, the angiosperms account for

A.   ANTHOXANTHINS

TYPE COMPOUND	DESCRIPTION OF CLASS
1. Flavones 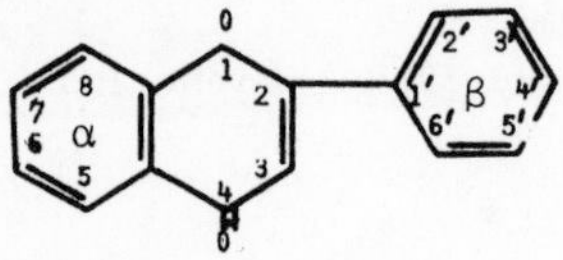 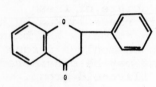	Hydroxylated and/or methoxylated derivatives of flavone The 3-hydroxyflavones are commonly referred to as flavonols.
2. Flavanones 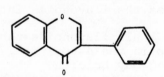	Hydroxylated and/or methoxylated derivatives of flavanone (2,3-dihydroflavone). The 3-hydroxyflavanones are commonly referred to as flavanonols.
3. Isoflavones	Analogous to the flavones with the aromatic ring linked to carbon 3 instead of carbon 2.
4. Chalcones 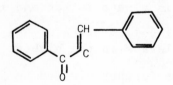	Hydroxylated and/or methoxylated derivatives of two aromatic rings linked by a three carbon aliphatic fragment.

Figure 1.   Classification of flavonoids

A.   ANTHOXANTHINS

TYPE COMPOUND	DESCRIPTION OF CLASS
5. Aurones	Hydroxylated and/or methoxylated derivatives of benzalcoumranone.

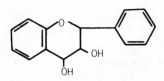

- -

B.   FLAVANS

Catechins — 3,7,4'-Hydroxyflavans which may also be hydroxylated at the 5, 3', and/or 5' positions.

Leucoanthocyanins — Hydroxylated and/or methoxylated derivatives of 3,4-dihydroxyflavan.

Figure 1. Classification of flavonoids (continued)

Cinnamic acids (3-phenylproponoic acid derivatives)

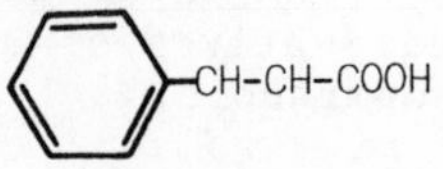

Quinic acid (1,3,4,5-tetrahydroxycyclohexanecarboxylic **acid**),

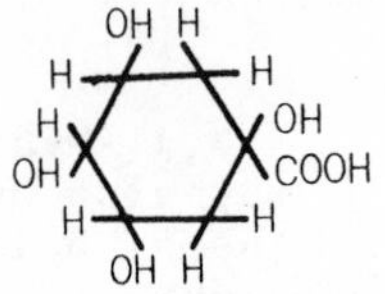

Quinic acid esters of cinnamic acids, and coumarins
(hydroxy and/or methoxy derivatives of coumarin).

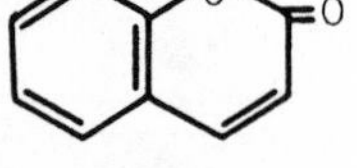

Figure 2.   *Classification of related compounds*

between 250 and 300 thousand species. Of this number
less than 400 species are cultivated or gathered as
human foods. These 400 species include 33 of the 51
orders and 89 of the 279 families. All parts of plants
are eaten - roots, stems, leaves, flowers, fruit, and
seeds - but in most species the edible portions are
restricted to one part. Flavonoids and related compounds have been isolated from, or detected in about
one-half of these edible plants, but not always in the
edible portions. The same compound, or group of compounds, are not always present throughout the plant.

Flavonoids occur in all types of higher plant
tissue - wood, bark, stems, leaves, fruit, roots,
flower, pollen and seeds. Table I shows the general
distribution of flavonoids and cinnamic acids in the
various parts of the plant. Certain groups of flavonoids are more characteristic of some tissues than
others. In fruit bearing plants, however, the same
groups of flavonoids that occur in the leaves also
occur in the fruit in a lesser amount. Anthocyanins
are typically in fruits, flowers, and some leaves.
The greatest natural source of flavans - catechins and
leucoanthocyanins - are from woods and barks. However, these do occur in non-woody tissue as tea leaves,
cocoa beans, and fruit pulps. Chalcones and aurones
are chiefly found in flower petals, and to a lesser
extent in leaves and stems of some species but are not
as widely distributed as other groups of flavonoid
compounds. Flavones and flavonones are present in
many plant tissues and cannot be considered as components of any one type of tissue.

Perhaps the greatest stimulus to the study of
flavonoids and related compounds came with the development of paper chromatography and its evolution into
thin-layer techniques. Paper chromatography provided
the first satisfactory procedures of surveying plant
tissues for the presence of flavonoids (4). These
compounds possess just the right range of solubility
characteristics for ease in separation (5, 6) and most
of them possess characteristic spectra in ultraviolet
and/or visible regions (6, 7).

Nearly thirty years ago paper chromatography was
employed for the separation of flavonoids. There have
been many excellent reviews on the subject (5, 6, 8,
9, 10, 11). The selection of a specific chromatographic
procedure depends on the objectives of an investigation.
The isolation and purification of a flavonoid (or a
cinnamic acid) can be achieved by various preparation
techniques using either one- or two-dimensional procedures. In preparative procedural work in our

laboratory has usually been by series of one-dimensional techniques. In either case, several separations are required and several detection techniques must be used.

Many of the flavonoids and related compounds have strong antioxidant characteristics in lipid-aqueous and lipid food systems (Tables II, III, Figure 3). As may be seen certain flavones, flavonols, flavonones, flavanonals, and cinnamic acid derivatives have considerable antioxidant activity. The very low solubility of these compounds in lipids is often considered a disadvantage and is considered a serious disadvantage if an aqueous phase is also present (12). However, flavonoids suspended in the aqueous phase of a lipid-aqueous system offer appreciable protection to lipid oxidation (13, 14, 15, 16, 17, 18). Also, Lea and Swoboda (12), nearly twenty years ago, found that flavonols were effective antioxidants when suspended in lipid systems.

The antioxidant activity was measured using 20 mg. of linoleic acid, 200 mg. of Tween 40, and 1 ml. of 0.02% B-carotene in chloroform. The chloroform was removed by evaporation on a water-bath at 50°C., using a rotary evaporator. 50 ml. of oxygenated water was added, and 5 ml. aliquots of this emulsion were placed in spectrometer tubes with 2 ml. of the antioxidant solution under test. For the control, 2 ml. of deionized, distilled water, or ethanol, as appropriate, were added to the emulsion. Readings at 470 nm. were taken immediately. The tubes were stoppered, and placed in a water-bath at 50°C. Readings of the optical density were taken at regular intervals until the control was bleached. The antioxidant index was calculated by dividing the loss of optical density of the control at the end of the induction period, by the loss of optical density of the test solution at that time.

Polyphenolic antioxidants, sparingly soluble in lipid systems, have been converted into readily fat-soluble form by alkylation or esterification with long chain fatty acids or alcohols. Such a procedure offers promising results with flavonoids.

The action flavonol antioxidation is bi-modal. Flavanols are known to form complexes with metals. Chelation occurs at the 3-hydroxy, 4-keto grouping and/or at the 5-hydroxy, 4-keto group, when the A ring is kydroxylated in the 5 position. An o-quinol grouping on the B-ring can also demonstrate metal-complexing activity (19, 20). However, the major value of flavonoids and cinnamic acids is in their primary

TABLE I. The General Distribution of Flavonoid Compounds in Plant Tissues

Plant Tissue	Relative Concentration of Compounds
Wood	Catechins ~ Leucoanthocyanins > flavonols > cinnamic acids
Bark	As wood but greater total quantity
Leaf	Flavonols ~ cinnamic acids > catechins ~ Leucoanthocyanins
Fruit	Cinnamic acids > catechins ~ Leucoanthocyanins > flavonols

TABLE II. Antioxidant Activity of Flavones

Compound	Antioxidant Index (5×10^{-4} M)
Aglycones:	
Quercetin (3,5,7,3',4'-Pentahydroxy)	3.8
Fisetin (3,7,3',4'-Tetrahydroxy)	3.8
Myricetin (3,5,7,3',4',5'-Hexahydroxy)	4.5
Robinetin (3,7,3',4',5'-Pentahydroxy)	4.5
Rhammnetin (3,5,3',4'-Tetrahydroxy 7-Methoxy)	3.6
Glycosides:	
Quercetin (Quercetin 3-Rhamnoside)	3.7
Rutin (Quercetin 3-Rhamnoglucoside)	1.6

TABLE III. Antioxidant Activity of Flavanones

Compound	Antioxidant Index (5×10^{-4} M)
Aglycones:	
Naringenin (5,7,3'-Trihydroxy)	1.6
Dihydroquercetin (3,5,7,3',4'-Pentahydroxy)	3.8
Hesperitin (5,7,3'-Trihydroxy-4'-Methoxy)	1.2
Glycosides:	
Hesperidin (Hesperitin 7-Rhamnoglucoside)	1.2
Neohesperidin (Hesperitin 7-glucoside)	1.3

Compound	Structure	Antioxidant Index (5×10^{-4}M)
Hesperidin Methyl Chalcone	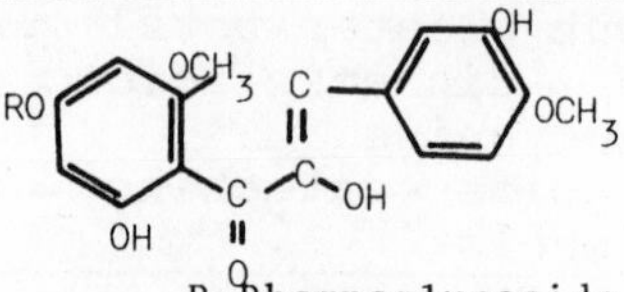 R-Rhamnoglucoside	1.3
D-Catechin	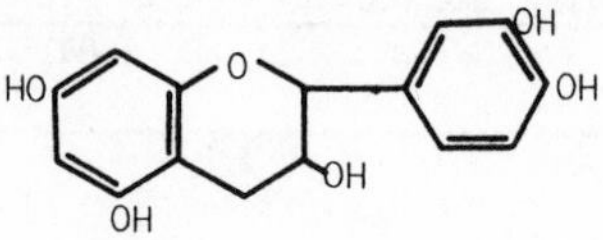	3.5
Chlorogenic Acid	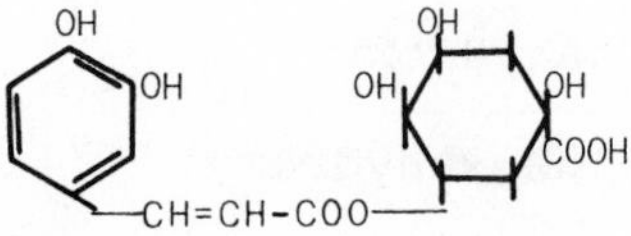	3.7
Caffeic Acid	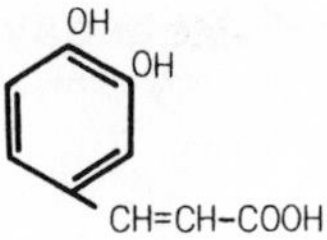	3.6
Quinic Acid	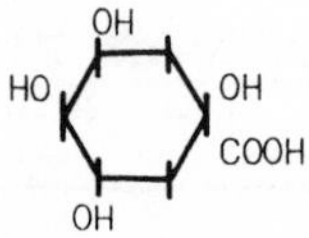	1.5
Propyl Gallate	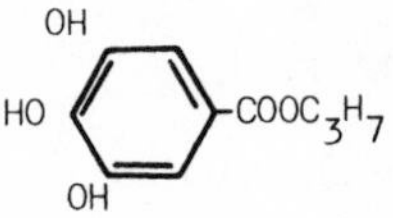	2.1

Figure 3. Antioxidant indices of some flavonoids and related compounds

antioxidant activity (i.e., as free radical acceptors
and as chain-breakers).

The major evidence that these compounds work
mainly as primary antioxidants is their ability to
work equally well in metal cayalyzed and uncatalyzed
systems. They are also efficient antioxidants in
systems catalyzed by relatively large molecules, such
as heme and other porphyin compounds. They are also
effective against lipoxygenase catalyzed reactions.
These compounds cannot be enfisaged as forming complexes with flavonols. In addition, hesperitin (5,7,3' -
trihydroxy-4'methoxyflavone) which possesses an active
metal-complexing site has demonstrated neglibible
antioxidant activity.

The position and the degree of hydroxylation is
of primary importance in determining antioxidant activity. There is general agreement that ortho-dihydroxyl-
ation of the B ring contributes markedly to the antioxidant activity of flavonoids (12, 13, 14, 21, 22, 23,
24). The para-quinol structure of the B ring has been
shown to impart even greater activity than the ortho-
quinol structure; while the meta configuration has no
effect on antioxidant activity (21). However, para
and meta hydroxylation of the B ring apparently does
not occur commonly in nature.

All flavonoids with the 3',4'-dihydroxy configuration possess antioxidant activity. Two (robinetin
and myricetin) have an additional hydroxyl group at
the 5' position, which increases the antioxidant acti-
vities over those of the corresponding flavones with
the 5'-hydroxyl group, fisetin and quercetin. Two
flavanones (naringenin and hesparitin) having a single
hydroxyl group on the B ring possesses only slight
antioxidant activity. Hydroxylation of the B ring is
a major consideration for antioxidant activity.

Meta 5,7-hydroxylation of the A ring apparently
has little, if any, effect on antioxidant activity.
This is efidenced by the findings that quercetin and
fisetin have relatively the same activity and myricetin
possesses the same activity as robinetin. Heimann and
his associates (23, 24) reported that meta 5,7-hydroxyl-
ation lowered antioxidant activity. To the contrary,
Mehta and Seshadri (22) found quercetin to be a more
effective antioxidant than 3,3',4'-trihydroxyflavone.
Data from our laboratory support the finding of Mehta
and Seshadri.

The importance of other sites of hydroxylation
were studied by Lea and Swoboda (12); Mehta and Seshadri
(22); Simpson and Uri (21); and Uri (25). The two
former groups found quercetagetin (3,4,5,7,3,4'-hexa-

hydroxyflavone) and gossypetin (3,5,7,8,3',4'-hexa-
hydroxyflavone) to be very effective antioxidants.
Uri (25) found that the ortho-dihydroxy grouping on
one ring and the paradihydroxy grouping on the other
(i.e., 3,5,8,3'4'- and 3,7,8,2',5'-pentahydroxy-
vlavones) produced very potent antioxidants. These
four polyhydroxyflavones are the most potent flavonoids,
as antioxidants, yet reported in non-aqueous systems.
Simpson and Uri (21) found 7-n-butoxy-3,2',5'-trihy-
droxyflavone to be the most effective antioxidant of
30 flavones studied in aqueous emulsions of methyl
linoleate.

The 3 glycosides possess approximately the same
antioxidant activity as the corresponding aglycone
when the glycosyl substitution is with monosaccharide.
In the case of rutin where the substitution is with
a disaccharide antioxidant activity is reduced. The
antioxidant capacity of a commercial preparation of
rutin is considerably lower than the corresponding
aglycone, quercetin. Kelley and Watts (19) studied
the antioxidant effect of several flavonoids and found
rutin womewhat inferior to quercetin and quercitrin
but the differences were not as great as we have found.
Chromatographic purification and the use of several
commercially available samples (to eliminate the effect
of possible contamination) did not alter the finding.
Kelley and Watts (19), using a carotene-lard system
also found that quercitrin had approximately the same
protection as quercetin. Crawford et al, (26) found
that methylation of the 3-hydroxyl group of quercetin
only slightly lowered antioxidant activity. However,
considerable importance has been attached to the free
3-hydroxyl by others (12, 21, 22, 23). Mehta and
Seshadri (22) postulated that the 3-hydroxyl and the
2,3 double bond allowed the molecule to undergo isomeric changes to diketo forms which would possess a
highly reactive-CH group (position 2).

Dihydroquercetin was found to have the same antioxidant activity as quercetin indicating either that
the 2,3, double bond is not of major importance to
antioxidant activity or that conversion of dihydroquercetin to quercetin took place while the compound
was in contact with the oxidizing fat. Mehta and
Seshadri (22) suggested that conversion might account
for the antioxidant activity of dihydroquercetin. However, chromatographic tests demonstrated that dihydroquercetin is not converted to quercetin by the hydrolysis procedure, nor could quercetin be chromatographically detected in the carotene-lard system in which
dihydroquercetin was used as antioxidant. Dihydro-

quercetin was still present after 12 hours in the system.

Perhaps the greatest potential source of flavonoids for food antioxidants is from wood as a by-product of lumber and pulping operations. Whole bark of the Douglas fir contains about five percent dihydroquercetin (3,4,7,3',4 pentohydroxyflavonone). The cork fraction, readily separated from the bark, contain up to 22% dihydroquercetin (27). Kirth (28) reported that approximately 150 million pounds of dihydroquercetin are potentially available annually in Oregon and Washington alone. Quercetin (3,5,7,3'4' pentohydroxyflavone) has been produced commercially as an antioxidant from wood sources (29). Quercetin is present in much lower amounts in wood and bark than is dihydroquercetin but quercetin can be obtained in quantity by oxidation of dihydroquercetin.

As mentioned earlier other plant constituents which might be expected to show antioxidant powers would be primarily phenolic compounds, expecially o- and p- dihydroxy phenols such as the hydroxy cinnamic acids, caffeic and ferulic acids. While these acids usually occur in plant tissue as water-soluble esters, commonly chlorogenic acid or daffeoyl-quinic acid, and sugar esters, they have also been isolated as complex lipophilic esters of glycerol, long-chain diols, and ω-hydroxy acids. These lipophilic esters have been revealed as antioxidants in a comprehensive investigation of the antioxidants in oats (30, 31, 32, 33). The caffeoyl esters have considerably more antioxidant activity than do those of ferrulic acid. Other lipid-soluble esters of ferulic acid with cycloartenol and other triterpenoids have been shown by Ohta et al. (34) to occur in rice bran oil, while a ferulate of dihydroxy-B-sitosterol has been isolated from maize by Tamura et al. (35). Wheat has also been shown to contain similar steroid esters (36).

The presence of two isomers of chlorogenic acid, also ferulic acid, and several other phenolic acids, has been confirmed in hexane defatted soy flour by Arai et al. (37).

Literature Cited

1. Geissman, T. A. and Hinreiner, E. Botan. Rev. (1952) 18:77.
2. Bate-Smith, E. C. Advances in Food Research. (1954) 5:261.
3. Geissman, T. A.   "The Chemistry of Flavonoid Com-

pounds", Chapter 1.   (Edited by T. A. Geissman)
The Macmillan Company, New York.   (1962).
4. Bate-Smith, E. C. Chem. Ind. R.   (1956)   32.
5. Harborne, J. B. J. Chromatography. (1959) 2:581.
6. Mabry, T. J., Markham, K. R. and Thomas, M. B.
"The Systematic Identification of Flavonoids".
Springer-Verlag, Berlin   (1970).
7. Jurd, L. "The Chemistry of Flavonoid Compounds",
Chapter 5.   (Edited by T. A. Geismann) The Macmillan Company, New York.   (1962).
8. Harborne, J. B. J. Chromatography (1958) 1:473.
9. Thompson, J. F., Honda, S. I., Hunt, G. E., Krupka,
R. M., Morris, C. J., Powell, Jr., L. E., Silber-
stein, O. O., Towers, G. H. N., and Zacharius,
R. M. Botan. Rev. (1959) 25:1.
10. Seikel, M. K. "The Chemistry of Flavonoid Compounds", Chapter 3 (Edited by T. A. Geissman)
The Macmillan Company, New York.   (1962).
11. Seikel, M. K. "Biochemistry of Phenolic Compounds",
Chapter 2 (Edited by J. B. Harborne) Academic
Press, New York.   (1964).
12. Lea, C. H. and Swoboda, P. A. T. Chem. Ind.   (1956)
1426.
13. Pratt, D. E. and Watts, B. M. J. Food Sci.   (1964)
29:27.
14. Pratt, D. E. J. Food Sci. (1965) 30:737.
15. Cofer, A. Unpublished data. (1961) Florida
State University.
16. Cofer, A. Unpublished data. (1963) Florida
State University.
17. Cofer, A. Ph.D. Thesis. (1965) Florida State
University.
18. Ramsey, M. B. and Watts, B. M. Food Technol.
(1963) 17:1056.
19. Kelley, G. G. and Watts, B. M. Food Research
(1957) 22:308.
20. DeWitt, K. W. Chem. Ind. (1955) 1551.
21. Simpson, T. H. and Uri, N. Chem. Ind. (1956)
22. Mehta, A. C. and Seshadri, T. R. J. Sci. Ind.
Research (1959) 18B:24.
23. Heimann, W., Heimann, A., Gremminger, M. and Holman,
H. H. Fette u. Soifen (1953) 55:394.
24. Heimann, W. and Reiff, F. Fette u. Soifen (1953)
55:451.
25. Uri, N. 1961. Mechanism of antioxidation.
"Autoxidation andAntioxidants", Chapter 4
(Edited by W. O. Lundbert) Interscience Publishers,
New York.
26. Crawford, D. L., Sinnhuber, R. O. and Aft. H.
J. Food Sci. (1962) 26:139.

27. Hergert, H. L. and Kurth, E. F. Tappi (1952)
    35:59.
28. Kirth, E. F. Ind. Eng. Chem: (1953) 45:2096.
29. Anon. Chem. Eng. News (1958) 36 (No. 7), 58.
30. Daniels, D. G. H., King, H. G. C. and Martin, H. F.
    J. Sci. Food Agr. (1963) 14:385.
31. Daniels, D. G. H. and Martin, H. F. Chem. Ind.
    (1964) 2058.
32. Daniels, D. G. H. and Martin, H. F. J. Sci. Food
    Agr. (1967) 18:589.
33. Daniels, D. G. H. and Martin, H. F. J. Sci. Food
    Agr. (1968) 19:710.
34. Ohta, G. and Shimuzu, M. Pharm. Bull. (Tokyo)
    (1957) 5:40.
35. Tamura, T., Sakaedani, N. and Matsumoto, T.
    Nippon Kagaku Zasshi (1958) 29:1011.
36. Tamura, T., Hibino, T., Yokoyama, D. and Matsumoto,
    T. Nippon Kagaku Zasshi (1959) 80:215.
37. Arai, S., Suzuki, H., Fujimaki, M. and Sakurai, Y.
    Agr. Biol. Chem. (Tokyo) (1966) 30:364.

2

Contribution of Polyphenolic Compounds to the Taste of Tea

GARY W. SANDERSON, ARVIND S. RANADIVE, LARRY S. EISENBERG,
FRANCIS J. FARRELL, ROBERT SIMONS, CHARLES H. MANLEY, and
PHILIP COGGON

Thomas J. Lipton, Inc., 800 Sylvan Ave., Englewood Cliffs, N. J. 07632

What Is Tea?

Tea is a processed vegetable material used to prepare a
stimulating, delicately flavored beverage that is one of the
most popular drinks in the world. Tea is manufactured from the
tender shoot tips (i.e. the "flush") of the tea plant Camellia
sinensis, (L.) O. Kuntze, cultivated in many tropical and sub-
tropical areas around the world. The tea manufacturing process (1,2,3) causes the fresh green tea leaf to be converted
to commercial tea products such as green tea (not fermented),
oolong tea (partially fermented), or black tea (fully fermented).
Tea fermentation refers to an oxidation of the flavanols found
in the tea leaf which is brought about by a catechol oxidase
enzyme that is endogenous to the leaves of tea plants (2,4):
Control of this reaction is central to good tea manufacturing
practices.

The chemical composition of a tea beverage prepared from a
commercial black tea blend, i.e. a Lipton tea bag, is shown in
Table 1. This set of analyses agrees closely with other analyses that have been published (5,6,7) indicating that the black
tea studied in this investigation is representative of black
teas in general.

As shown in Table 1, polyphenolic compounds are estimated to
comprise about 48.5% of the total solids in a cup of tea. As will
be shown later in this paper, these polyphenolic compounds make
a most important contribution to the taste of tea, and the exact nature of this contribution is determined by the kind of
polyphenolic compounds that are present in the tea beverage.
Accordingly, to understand the chemistry underlying the taste of
tea, one must understand the chemistry of tea manufacture, especially the tea fermentation process, since this determines the
makeup of the polyphenolic compounds in tea products.

Returning for a moment to our question; namely, "What is
tea?", we recognize that most people think of tea as the beverage that they obtain by steeping a tea bag containing black tea

14

in boiling hot water, or by dissolving instant tea in cold water.
In either case, the average tea beverage in the United States has
a tea solids concentration of about 0.30% tea solids obtained by
steeping a tea bag (contains about 2.27g black tea leaf) in a cup
with about 6 oz. of hot water (initially at about 100°C) for
about 1 min. (this produces about 5.2 oz. of beverage after the
tea bag is removed), or by dissolving a gently rounded teaspoon-
ful of instant tea (about 0.70g of instant tea solids) in 8 oz.
of cold water. Most of our studies have been carried out on an
approximation of this standard American black tea beverage prepared from tea bags (also called a tea infusion) since it is of
greatest concern to the authors.

    As will be explained later, the caffeine in tea has an im-
portant modifying effect on the taste of tea beverages. Accordingly, it is noteworthy that the caffeine concentration in the
black tea beverage studied in this investigation (Table 1) was
0.026% (8.61% of the tea extract solids themselves) which equates
to about 40mg caffeine in the "average cup of tea". This value
for the amount of caffeine in a cup of tea compares with the value
of 41 mg/cup reported by Burg (8) who investigated this matter.

The Chemistry Of Tea Manufacture

    The chemistry of tea manufacture is described in some de-
tail elsewhere (2,9). In brief, one begins the black tea manufacturing process by plucking (harvesting) the flush of the rapidly growing tea plants. The flush is particularly rich in polyphenolic compounds (2,9) and of particular importance are the
flavanols, i.e. (-)-epicatechin (I), (-)-epicatechin-3-gallate
(II), (-)-epigallocatechin (III), (-)-epigallocatechin-3-gallate
(IV), (+)-catechin (V), and (+)-gallocatechin (VI). The total
amount of flavanols present in fresh tea flush will vary from
about 15 to 25% (dry weight basis): The exact amount of these
compounds present in any particular lot of freshly harvested
tea shoot tips is determined by horticultural factors such as
the clones of tea plants from which the tea shoot tips were harvested and the climate that prevailed while the tea shoot tips
were developing.

    The tea manufacturing process begins within a few hours after
harvesting of the fresh tea flush, and the fermentation step is
the most characteristic, and the most important, step in the process. Tea fermentation is initiated by macerating the fresh tea
shoot tips causing the endogenous tea catechol oxidase to come
into contact with the flavanols that are also present in these
tissues. The consequence of this process is an oxidation of the
flavanols (I-VI), and gallic acid (VIIa) by coupled oxidation
(10), which leads to the formation of the bis-flavanols A (VIII),
B (IX), and C (X); theaflavin (XI); theaflavin gallates A (XII)
and B (XIII); theaflavin digallate (XIV); epitheaflavic acid (XV);
3'-galloyl-epitheaflavic acid (XVI); and thearubigins which are

Table 1: Proximate Analysis of a Black Tea Infusion. (The black tea used was Lipton tea bag blend and the tea extract solids represented about 33% of the total tea leaf dry weight. See the Experimental Section for more details on analytical procedures).

Chemical Constituent	Amount of Each Constituent	
	In the Beverage, Calcd (% x 10^2)	In the Tea Extract Solids (%)
Water	9969	4.75
Polyphenols, total	16.0	48.5
(-)-Epicatechin (I)	0.4	1.15
(-)-Epicatechin-3-gallate (II)	1.2	3.68
(-)-Epigallocatechin (III)	0.3	1.04
(-)-Epigallocatechin-3-gallate (IV)	1.5	4.41
Flavonol glycosides and others	Trace	Trace
Bisflavanols (VIII-X)	Trace	Trace
Theaflavins (XI-XIV)	0.8	2.50
Epitheaflavic acids (XV-XVI)	Trace	Trace
Thearubigins (XVII-XX and other unknowns)	11.4	34.2
Gallic acid (VIIa)	0.4	1.1
Chlorogenic acid	0.1	0.2
Caffeine	2.4	7.20
Theobromine	0.1	0.24
Theophylline	0.2	0.66
Carbohydrates		
Polysaccharides, total (by difference)	1.3	3.97
Pectin	0.05	0.15
Sugars, total	2.2	6.52
Fructose	0.60	2.0
Glucose	0.57	1.9
meso-Inositol	0.15	0.50
Sucrose	0.48	1.6
Maltose	0.03	0.1
Raffinose	0.10	0.3

(continued on next page)

Table 1, (continued)

Chemical Constituent	Amount of Each Constituent		
	In the Beverage, Calcd (% x 10^2)		In the Tea Extract Solids (%)
Organic Acids, total	0.8		2.52
(pH)		(5.1)	---
(Total acidity as citrate)		(2.36)	(7.85)
Oxalic		0.42	1.4
Malonic		0.01	0.02
Succinic		0.02	0.09
Malic		0.09	0.30
trans-Aconitic		0.003	0.01
Citric		0.27	0.80
Lipids, total	1.5		4.56
Minerals (Ash)	3.0		9.08
Potassium		1.37	4.6
Sodium		0.03	0.11
Calcium		0.02	0.08
Magnesium		0.075	0.25
Iron		0.0013	0.005
Maganese		0.015	0.05
Aluminum		0.014	0.05
Peptides (6.25 x N)[a]	1.9		5.71
Amino acids, total	2.1		6.29
Aspartic acid		Trace	0.39
Threonine		Trace	0.07
Serine		Trace	0.24
Glutamic acid		Trace	0.42
Glycine		Trace	0.02
Alanine		Trace	0.12
Valine		Trace	0.20
Methionine		Trace	0.02
Isoleucine		Trace	0.18
Leucine		Trace	0.19
Tyrosine		Trace	0.15
Phenylalanine		Trace	0.16
Ammonia		Trace	0.13
Lysine		Trace	0.04
Histidine		Trace	0.002
Arginine		Trace	0.03
Glutamine		Trace	0.19
Asparagine		Trace	0.24
Tryptophane		Trace	0.09
Theanine		1.1	3.40

[a] Total non-caffeine N (1.92%) x 6.25 — amino acids

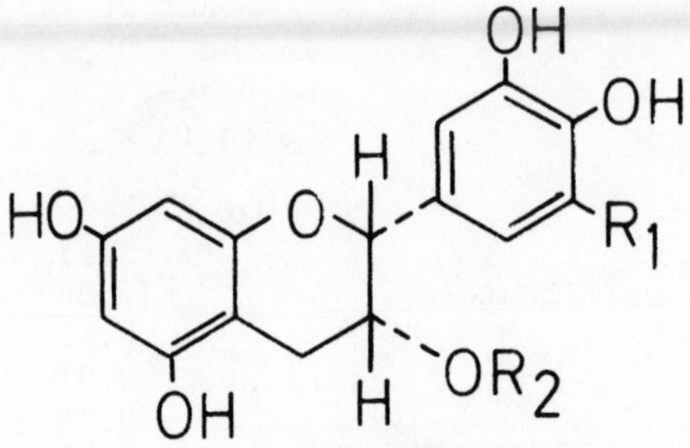

I,   (-)-Epicatechin; R_1=H; R_2=H
II, (-)-Epicatechin-3-gallate; R_1=H; R_2=VIIb
III, (-)-Epigallocatechin; R_1=OH; R_2=H
IV, (-)-Epigallocatechin-3-gallate; R_1=OH; R_2=VIIb

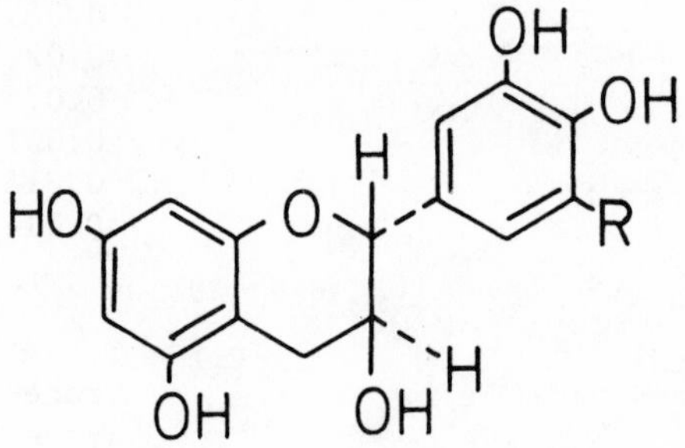

V, (+)-Catechin; R=H
VI, (+)-Gallocatechin; R=OH

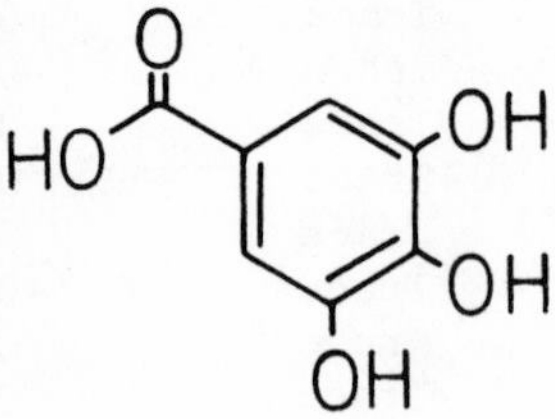

VIIa, Gallic acid

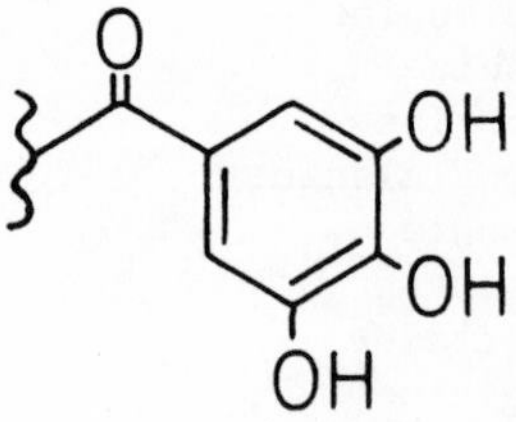

VIIb, Galloyl group

VIII, Bisflavanol A; R_1=VIIb; R_2=VIIb
IX, Bisflavanol B; R_1=H; R_2=VIIb
X, Bisflavanol C; R_1=H; R_2=H

XI, Theaflavin; R_1=H; R_2=H
XII, Theaflavin gallate A; R_1=H; R_2=VIIb
XIII, Theaflavin gallate B; R_1=VIIb; R_2=H
XIV, Theaflavin digallate; R_1=VIIb; R_2=VIIb

polymeric proanthocyanidins; i.e. procyanidin (XVII), procyanidin-3-gallate (XVIII), prodelphinidin (XIX), prodelphinidin-3-gallate (XX). The chemistry of the thearubigins is only poorly understood at the present time, but it is known that they are a heterogeneous group of polymers formed by the oxidative condensation of the simple flavanols (I-VI) (11). Further, the thearubigins have been characterized as polymeric proanthocyanidins (12, 13, 14), with molecular weights ranging from 700-40,000 (7). Finally, the exact composition of the thearubigins probably varies with the conditions of their formation; i.e. the conditions of black tea manufacture, which has made their determination a most difficult matter (15). The reactions of the tea polyphenols during tea fermentation and firing are outlined in Figure 1. Fermentation and firing leads to the insolubilization of the tea leaf proteins, some of the tea polyphenolic material, and other substances, but considerable solid matter remains extractable with boiling water. Those black tea substances that are extracted in the "normal" brewing of black tea leaf are listed in Table I.

In green tea manufacture, the harvested tea shoot tips are steamed prior to maceration in order to inactivate the endogenous catechol oxidase enzyme. As a result, the tea flavanols undergo very little change in this process and green tea is rich in unoxidized flavanols (I-VI). In black tea manufacture, the tea fermentation process is allowed to proceed to near completion so there are usually only traces of unoxidized flavanols (I-VI) remaining in the finished product. However, the exact mix of flavanol oxidation products (VIII-XX, and others) will vary depending on the precise conditions under which the tea manufacturing process takes place. Oolong tea (commonly called Chinese tea in the United States) is produced by a partial fermentation so it contains an appreciable residue of unoxidized flavanols (I-VI) as well as the flavanol oxidation products (VIII-XX and others).

In addition to the polyphenolic compounds, aroma is most important in determining the flavor and quality of tea products. Many chemical changes take place during the tea manufacturing process, especially during tea fermentation and the subsequent firing (drying) step, that are essential to the formation of the aroma characteristic of tea. It has been shown (16) that the oxidation of the tea flavanols that takes place during tea fermentation is itself an essential driving force for reactions that are required to develop the aroma that is characteristic of tea products, and firing has been shown to be essential for black tea aroma formation (17, 18). The chemistry of flavor formation during the manufacture of black tea was recently reviewed (19, 20), and it is summarized in Figure 2.

Attention should be drawn to caffeine (XXI) since caffeine does play a part in determining the taste of a cup of tea. Caffeine is biosynthesized in the tea plant (21, 22), and it undergoes practically no change during the black tea manufacturing process (2). Therefore, the amount of caffeine present in tea prod-

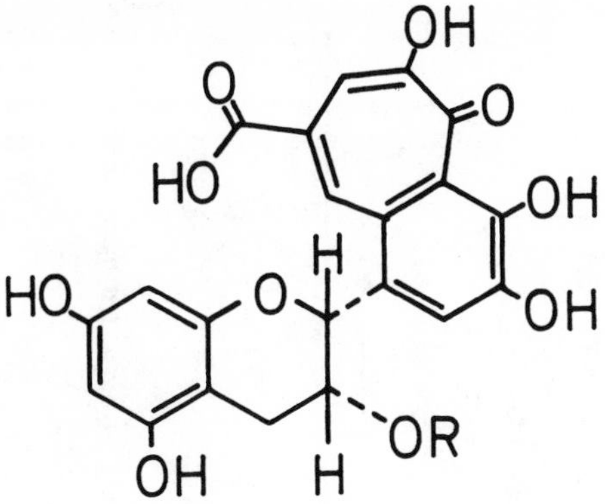

XV, Epitheaflavid acid;
   R=H
XVI, 3'-Galloyl epitheaflavic
   acid; R=VIIb

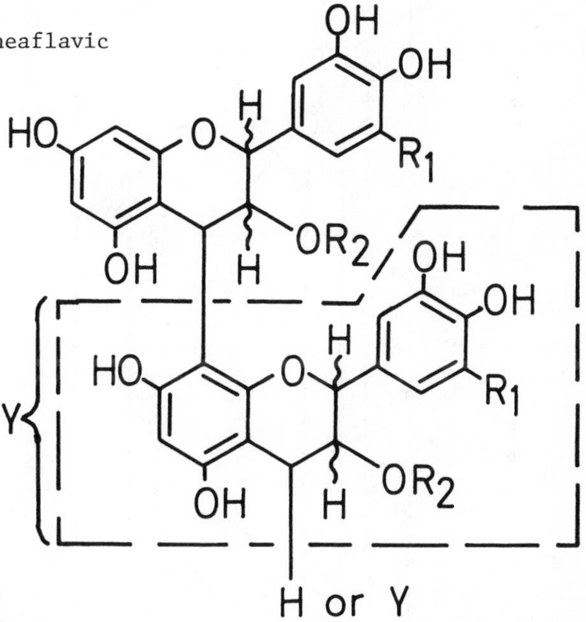

H or Y

XVII, Procyanidin; R_1=H; R_2=H
XVIII, Procyanidin gallate,
   R_1=H; R_2=VIIb

XIX, Prodelphinidin; R_1=OH; R_2=H
XX, Prodelphinidin gallate;
   R_1=OH; R_2=VIIb

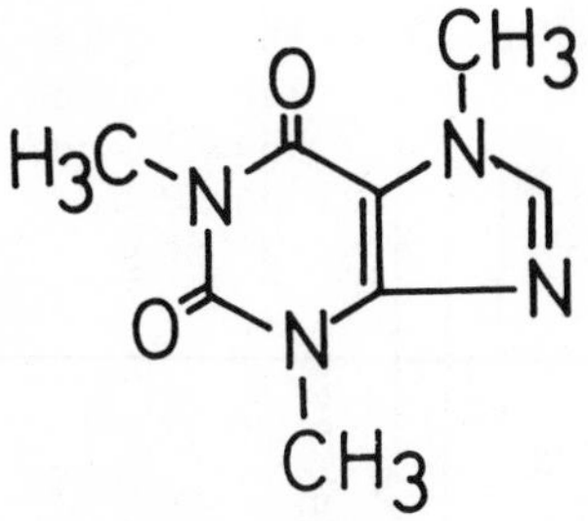

XXI, Caffeine

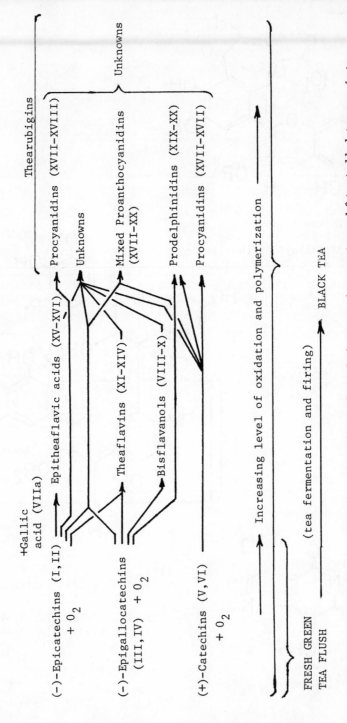

Figure 1. Summary of changes undergone by the tea flavanols during tea fermentation and firing in black tea manufacture

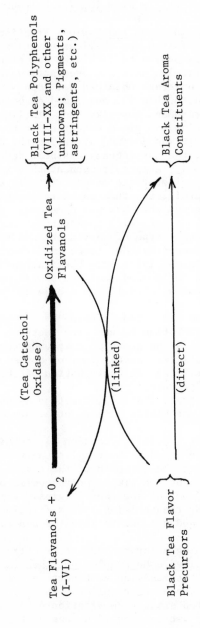

Figure 2. *Summary of reactions taking place during tea fermentation and firing in black tea manufacture. Tea flavanol oxidation has an essential role in causing various chemical changes that are important to the formation of black tea flavor*

ucts is fixed by horticultural practices, and cannot be changed
by currently known tea manufacturing processes.

The Taste of Black Tea Derives Mainly From The Tea Polyphenols

The Taste of a Black Tea Infusion and Various Fractions of
This Infusion. A black tea infusion was chemically analyzed
(Table 1) and organoleptically evaluated (Table 2, Fraction 1) as
a first step in our program to elucidate the chemistry underlying
the taste of black tea. The beverage obtained was found to have
a characteristic black tea taste that was described as being
"flowery, pleasing, mildly green and hay-like, and distinctly
black tea like". When attention was given specifically to the
astringency of the infusion, it was decided after lengthy delib-
eration by the panelists that the astringency was best described
as having two components; namely, a tangy component that was
sharp and puckery with little after taste effect (this is a difficult to describe type of astringency that is characteristic of
black tea), and a non-tangy component that was completely tasteless, mouth-drying and mouth coating, with a lingering (more than
60 sec.) after taste effect (this type of astringency is typical
of unripe bananas). It is noteworthy that there is virtually no
bitterness in this whole black tea infusion. Chemical analysis
of the "whole black tea infusion" showed that the solids were
composed of about 48% polyphenols, 7% caffeine and 44.3% "other"
materials (Table 1).
  Next, we fractionated the black tea infusion by a combination
of solvent extractions and adsorption column treatments in antici-
pation of being able to identify the group of compounds responsible for each component of black tea taste. And, of course, we
were most interested in identifying the contribution of the polyphenols to the taste of this product.
  The trichloroethylene extract (Table 2, Fraction 2) contained
mostly caffeine (XXI) and it was bitter with no other noticeable
taste attributes.
  The ethyl acetate extract (Table 2, Fraction 4) contained
mostly neutral black tea polyphenols. These polyphenols were
found by paper chromatography to be composed of the traces of unoxidized tea flavanols (I-VI) and the simple polyphenolic oxidation products (VIII-XVI) and some of the thearubigins: Roberts
et al. (23) named these the S_1 thearubigins. This fraction had
a trace of tangy astringency and a moderate level of non-tangy
astringency.
  The aqueous phase remaining after removal of Fractions 2 and
4 (Table 2, Fraction 5) contained some complex polyphenolic compounds named the S_{IA} and S_{II} thearubigins by Roberts et al (23),
and all the non-polyphenolic tea extract solids (Table 1) except
caffeine. This fraction tasted similar to Fraction 4 in that it
had a fair level of non-tangy astringency with none of the tangy
astringency.

Complete removal of polyphenolic compounds from the whole black tea extract (Fraction 1) was accomplished by passing this extract through a polyamide column (Table 2, Fraction 6). The removal of the polyphenolic compounds from the tea extract was accompanied by a removal of all astringency from the whole tea extract with the concomitant appearance of bitterness that was not present on the whole tea extract. These results, together with the results obtained for Fractions 4 and 5, indicate that the polyphenolic compounds in a whole black tea extract are astringent and that this astringency is expressed in the whole black tea extract. On the other hand, the caffeine in a black tea extract (Fraction 2) is present at a high enough concentration to produce a bitter taste, but this bitterness is not expressed in the whole black tea extract (Fraction 1); it is only expressed when the polyphenols are removed from the extract (Fraction 6). Removal of the caffeine from a polyphenol-free black tea extract (Fraction 6 → Fraction 7) was effective in removing the bitterness from this extract showing again that the caffeine in a black tea extract is responsible for a bitter taste in the absence of the black tea polyphenols.

Fraction 8 (Table 2) was prepared by addition of pure caffeine in the amount originally present in the whole black tea infusion (Fraction 1) to the neutral black tea polyphenols (Fraction 4). This caused a modest increase in the tangy astringency of Fraction 4 with no change in the non-tangy astringency. Most important, there was no noticeable bitterness in Fraction 8.

Fraction 9 (Table 2) was prepared by adding caffeine to Fraction 5 (i.e. the acidic black tea polyphenols and non-phenolic solids). This caused a small but significant decrease in the non-tangy astringency of Fraction 5, but no appearance of tangy astringency and no appearance of bitterness.

Fraction 10 (Table 2) is virtually a reconstitution of the whole black tea extract (Fraction 1), and, as might be expected, it was found to have taste properties that were very similar to the whole black tea extract.

Collectively, these results (Table 2) indicated that the black tea polyphenols are a central essential element in determining the taste of black tea infusions. This is illustrated best by noticing Fraction 6 which is virtually the complete black tea infusion minus the black tea polyphenols and which has practically no taste other than some bitterness: The bitterness is accounted for by the caffeine present in this fraction. The primary contribution of the black tea polyphenols to the "taste" of black tea infusions appears to be astringency, and the "tangy" portion of this astringency was found to be most characteristic of, and important to, the taste of black tea.

The results (Table 2) also suggest that caffeine plays a most important role in determining the level of tangy astringency in a black tea infusion. This was shown in two ways. First, decaffeination of a black tea infusion (i.e. Fraction 1 → Fraction 3)

Table 2. Composition and Taste of Fractions of Black Tea. (The
carried out as described in the Experimental Section.)

Fraction of Tea Extract	Composition of Fraction	
	Total Solids Present (%)	Amount of Poly-phenols (%)
(1) Whole black tea extract (Detailed composition shown in Table 1)	100[b]	48.5
(2) Trichloroethylene solubles (Mainly caffeine)	8.1	0
(3) Trichloroethylene extracted solids (Decaffeinated and dearomatized tea infusion)	92.3	48.5
(4) Ethyl acetate solubles (Neutral black tea polyphenols; i.e. flavanols, theaflavins, S_I thearubigins, etc.)	17.0	17
(5) Aqueous phase after removal of Fractions 2 and 4 (Acidic black tea polyphenols; i.e. S_{IA} and S_{II} thearubigins; and non-polyphenolic solids)	74.9	31.5
(6) Polyamide column effluent of Fraction 1 (Polyphenol free tea solids)	51.5	0
(7) XAD-2 column effluent of Fraction 5 (Polyphenol and caffeine free tea extract)	43.4	0
(8) Fraction 4 (Neutral black tea polyphenols) + caffeine	24.2	17
(9) Fraction 5 (Tea extract minus neutral black tea polyphenols) + caffeine	82.1	31.5
(10) Fraction 4 + Fraction 5 + Caffeine + Aroma	99.1	48.5

[a] Astringency ratings: 0 = none, 1 = threshold

[b] The total solids extracted were about 33% of

fractionation of the black tea beverage, Fraction 1, was

| | | Taste Description of Fraction | | |
| | | Astringency [a] | | |
Amount of Caffeine(%)	Polyphenol and Caffeine Free Solids (%)	Tangy	Non-Tangy	Other
7.2	44.3	3	3	Flowery, pleasing, mild green hay-like, black tea taste
7.2	Trace	0	0	Bitter
0.4	43.4	1	2	Very weak black tea taste
Trace	Trace	1	2	Sweetish after taste
Trace	43.4	0	3	Chalky
7.2	44.3	0	0	Slightly bitter, green hay like aroma
Trace	43.4	0	0	Malty
7.2	0	2	2	Plain
7.2	43.4	0	2	Chalky
7.2	43.4	2	3	Similar to Fraction 1

level, 2 = weak, 3 = moderate, 4 = strong.

the original Lipton black tea bag blend.

caused a marked reduction in the tangy astringency and in the black tea taste of the infusion. And second, while caffeine itself (Fraction 2) is bitter and has no astringency, the presence of caffeine together with the black tea polyphenols, and especially with the neutral black tea polyphenols, was necessary for the expression of a reasonable amount of tangy astringency (compare Fraction 4 and Fraction 8).

The aroma in the black tea infusion was also found to be important in determining the flavor of the beverages. None of the fractions (i.e. Table 2, Fractions 2-9) of the whole black tea infusion (Fraction 1) was noticeably black tea like unless aroma was present with the black tea polyphenols and caffeine (compare Fraction 10 with Fraction 1). It is noteworthy in this connection that aroma with caffeine and all other black tea solids except the polyphenols (Fraction 6) had a weak, slightly bitter, greenish taste that was not recognized as a black tea taste.

The Effect Of Tea Fermentation And Firing On The Taste Of Tea Infusions. Samples of tea were prepared that had been fermented for various periods of time and that had been either fired or not fired (i.e. frozen immediately after fermentation and freeze dried) for use in determining the effect of fermentation and firing on the tea polyphenols and on the taste of the resulting tea products. These samples were prepared in our laboratory using fresh tea flush, and it is recognized that the results of these experiments suffer from the limitations imposed by these non-optimal conditions. In spite of these limitations, we believe that our results are indicative of the role of tea polyphenols in determining the taste of black tea infusions.

The results of the analyses and the organoleptic evaluation of the infusions produced from these samples are summarized in Table 3. These results may be briefly summarized as follows:

(a) The amount of solids extracted from the tea leaf by the brewing process increases appreciably during the initial stage of tea fermentation, i.e. during the leaf maceration process and the very first minutes of the formal tea fermentation period, after which the extractable solids decrease as the tea fermentation period increases. Firing causes an additional appreciable loss of extractable solids. Apparently, the very first products of tea fermentation (Figure 1) are more easily extracted than the tea flavanols themselves, although tea fermentation (and firing) do lead to the formation of less and less soluble products as the length of the process increases.

(b) The amount of total flavanols in the infusion decreases as tea fermentation proceeds: This decrease in flavanols is most rapid in the early stages of tea fermentation. Firing causes an appreciable additional decrease in flavanols, especially in the initial stages of tea fermentation.

It is noteworthy that the gallo-flavanols (III,IV) are oxidized more rapidly than the catechol-flavanols (I,II) (Figure 3).

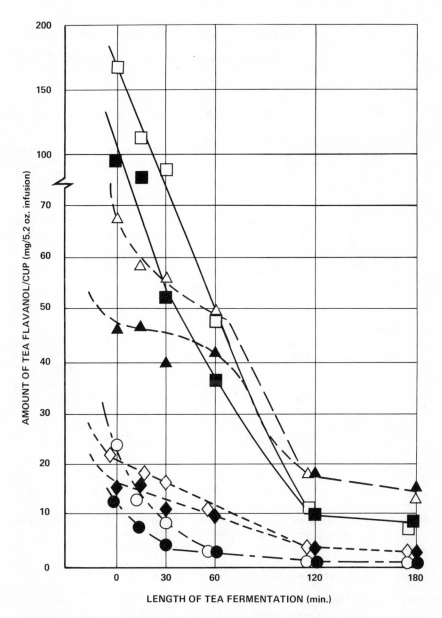

Figure 3.   The disappearance of tea flavanols in macerated tea flush as a result of tea fermentation and firing. Fermented and freeze dried (i.e. not fired): ◇, I and V; △, II; ○, III; □, IV. Fermented and fired: ◆, I and V; ▲, II; ●, III; ■, IV. These samples are described further in Table 3.

Table 3.   Effect of Tea Fermentation and Firing on the Taste of

Composition of Tea Infusion Solids [a]

Length of Tea Fermentation Period (min.)	Total Solids (mg/cup)	Total Flavanols (mg/cup)	Theaflavins (mg/cup)	Thea rubigins (mg/cup)
A. Fermented and then Freeze Dried				
0[c] (before maceration)	763	–	2.3	129
0 (after maceration)	872	283	4.1	202
15	837	206	9.4	219
30	804	171	12.7	232
60	824	111	17.3	274
90	825	–	16.3	278
120	781	34	13.1	276
180	732	26	13.6	283
240	746	–	12.2	294
B. Fermented and Fired				
0[c] (before maceration)	763	–	2.3	129
0 (after maceration)	851	173	7.4	242
15	828	154	10.1	256
30	786	107	12.9	271
60	758	92	14.7	285
90	735	–	15.4	293
120	706	33	13.3	290
180	680	29	11.9	276
240	630	–	11.5	276

[a] All tea infusions were prepared by extracting 2.27g of dry tea produced about 5.2 oz. of beverage (i.e. the infusion). Analysis caffeine/cup.

[b] Astringency ratings: 0 = none, 1 = threshold, 2 = weak,

[c] This sample of fresh green tea leaf was analyzed before any macerated prior to the formal tea fermentation period. Accord- fermentation, would have taken place in these samples prior to

Table 3 (cont.)

Tea Infusions

Organoleptic Evaluation of Infusions

Astringency [b]		Black Tea Aroma	Black Tea Taste	Other taste
Tangy	Non-Tangy			
0	1	0	None	Bland, slightly green
0	4	0	None	Very green, harsh, bitter
0	4	0	None	Very green, harsh, bitter
0	4	0	None	Very green, harsh, slightly bitter
0	3	0	None	Green, harsh
0	2	0	None	Green, slightly harsh
0	2	0	None	Green, slightly harsh
0	1	0	None	Green, slightly harsh
0	1	0	None	Green, slightly harsh
0	1	0	None	Bland, slightly green
0	4	0	None	Green, harsh, slightly bitter
0	4	0	None	Green, slightly harsh
0	3	0	None	Green, slightly harsh
1	2	1	Slight	Slightly green, slightly harsh
1	2	1	Mild	Slightly green, slightly harsh
1	2	1	Mild	Slightly harsh
1	1	2	Weak	None
1	1	2	Weak	None

leaf with 6.0 oz. boiling water in a tea cup for 5 min. This showed that each infusion contained between 38 and 40 mg.

3 = moderate, 4 = strong.

treatments. All other samples of tea leaf were withered and ingly, some oxidation of the tea leaf flavanols, i.e. tea the start of the formal tea fermentation period.

The reason for this differential in susceptibility to oxidation in the tea fermentation system is difficult to explain since all of the tea flavanols have similar affinities for the tea catechol oxidase enzyme (24), but the phenomena has been noticed before (18). Certainly, the amount of each flavanol, and the relative rate of oxidation of each flavanol, must be important determinants of the exact composition of their oxidation products (Figure 1), and consequentially on the taste of the finished black tea product.

(c) Theaflavins increase as the tea fermentation proceeds until a maximum is reached (after about 60 min. in the unfired samples and 90 min. in the fired samples) after which they decrease in amount. Firing causes a larger amount of theaflavins to accumulate after short fermentation periods and smaller amounts to be present after longer fermentation periods.

(d) An appreciable amount of thearubigins is formed in the macerated tea leaf prior to the start of the formal tea fermentation period: This must be due to the tea fermentation that takes place in injured cells of the tea flush during withering and, most important, the tea fermentation that takes place during maceration of the tea flush. The thearubigins continue to increase continuously as the tea fermentation period increases (Table 3, part A). Firing causes an additional appreciable increase in the amount of thearubigins extracted from samples with long, i.e. greater than about 120 min. in these experiments, fermentation periods (Table 3, part B). The decrease in extractable thearubigins after relatively long tea fermentation periods and firing appears to be closely related to the decrease in total extractable solids that is associated with these treatments.

(e) The tea fermentation that takes place prior to the formal tea fermentation period, i.e. during withering and maceration of the flush, causes a large increase in the non-tangy astringency of tea infusions prepared from this material. Tea fermentation per se (Table 3, part A) causes a decrease in non-tangy astringency that is present in withered, macerated tea leaf prior to the formal tea fermentation period, but neither tangy astringency nor black tea aroma develop, hence no black tea taste develops unless the samples undergo tea fermentation (for at least about 90 min. in these experiments) and are fired (Table 3, part B). It is noteworthy that the theaflavins and thearubigins content of the infusion from the unfired sample fermented for 180 min. was almost identical to the theaflavins and thearubigins content of the infusion from the fired sample that had been fermented for 120 min., yet these samples had entirely different taste profiles (i.e. the former was green and harsh with no black tea character whereas the latter was midly astringent and pleasantly black tea like): Very similar results and conclusions were reported previously (18). This points out the serious failings of the Roberts' method (15) for evaluating tea beverages by measurement of theaflavins and thearubigins in spite of much work to establish the validity of this test (25, 26, 27, 28).

The Taste Of Individual Polyphenolic Compounds Present In
Tea Infusions. The work described above established the importance of the tea polyphenols in determining the taste of black
tea infusions. Next we were interested in determining the taste
properties of each individual black tea polyphenol as a step
towards defining their separate contributions to the whole. Accordingly, the polyphenolic compounds present in tea beverages
were purified and then they were organoleptically evaluated for
their taste properties and their individual taste threshold values.
The results (Table 4) indicated that the only taste properties associated with the tea polyphenols are astringency and bitterness.
The simple, non-gallated tea flavanols (I, III, V) are not astringent, although they do have a bitter taste. On the other
hand, the simple, gallated tea flavanols (II, IV, VI) and the condensed tea flavanols (XI-XIX) are astringent in addition to having
a bitter taste. Of particular importance was the finding that in
no case was the astringency shown by the purified tea polyphenols
of the tangy type.

These results clearly show the importance of the galloyl
groups (VIIb) on the tea flavanols for the expression of astringency and bitterness. Results (Table 4) obtained with the various
theaflavins (XI-XIV) also indicates the importance of the galloyl
groups (VIIb) in determining the astringency of condensed (oxidized) polyphenolic compounds in tea. Theaflavin (XI) is formed
by oxidative condensation of (-)-epicatechin (I) and (-)-epigal-
locatechin (III) (which are not astringent), yet theaflavin (XI)
has some astringency even though it has no galloyl groups (VIIb):
This is presumably due to the relatively large molecular size
and the large number of phenolic groups of this molecule as compared to the simple non-gallated tea flavanols (I, III, V). However, there is a progressive increase in the intensity, i.e. decrease in the threshold level, of the astringency of the theaflavins as the number of galloyl groups (VIIb) per molecule increases. That is, theaflavin (XI) is less astringent that the
theaflavin monogallates A and B (XII-XIII) which are less astringent than theaflavin digallate (XIV).

The total of the unoxidized flavanols and the theaflavins in
black tea is hardly enough to reach their taste threshold level.
This leaves the thearubigins, which usually comprise over 30%
of all the black tea solids extracted into a cup of tea, to account for most of the "taste" of tea. Unfortunately, it was not
possible to purify the thearubigins sufficiently to determine
their threshold value, but it was determined that they are astringent. The chemistry of the thearubigins is only poorly understood at the present time, but it is known that they are a heterogeneous group of condensed flavans (11). It is noteworthy that
the astringency of tea beverages increases appreciably during the
very early stages of tea fermentation, i.e. before the beginning
of the formal tea fermentation period. The astringency of the
tea infusions then decreases steadily even though the level of

Table 4: Threshold Levels For Astringency and Bitterness Of Tea Polyphenols

Polyphenol	Threshold Level (mg/100ml)		Approximate Level in a Cup of Black Tea[a] (mg/100ml) of beverage)
	Astringency	Bitterness	
(-)-Epicatechin (I)	Not astringent	60	Trace
(-)-Epicatechin gallate (II)	50	20	Trace
(-)-Epigallocatechin (III)	Not astringent	35	Trace
(-)-Epigallocatechin gallate (IV)	60	30	16-18
(+)-Catechin (V)	Not astringent	60	Trace
Crude Theaflavins, a natural mixture (XI-XIV)	60	70	5-11
Theaflavin (XI)	80	75-100	0.6 - 1.2
Theaflavin monogallates A and B, a natural mixture (XII-XIII)	36	30- 50	1.8 - 3.7
Theaflavin digallate (XIV)	12.5	Not determined	2.4 - 4.8
Gallic acid (VIIa)	Not astringent[c]	Not bitter [c]	3-5
Thearubigins[b] (XVII-XX, others)	Not determined[d]	Not determined[d]	95-120
Tannic acid	20	80	0

[a]Based on amount of solids extracted from a standard American tea bag (2.27g. tea) brewed with 6 oz. boiling tap water in a cup for 3 min.
[b]Polymeric proanthocyanidins (12, 13).
[c]Tested at up to 1000 mg/100ml: Taste at this level was sour with sweet lingering after-taste.
[d]It was not possible to prepare samples of thearubigins of sufficient purity organoleptically to evaluate.

thearubigins continues to increase. It can be surmised that the astringency and the bitterness of the thearubigins derives both from the number of galloyl groups per individual thearubigin molecule and from the degree of condensation (size) of each individual thearubigin molecule which changes continuously during tea fermentation and firing, but these factors have yet to be studied. The lack of tangy astringency in any of the purified black tea polyphenols is thought to be related to the absence of caffeine in any of these purified preparations: Evidence for the importance of caffeine in determining black tea taste is given in other parts of this report (cf. Tables 2, 4 and 5).

The Effect Of Extraction Time On The Taste Of Black Tea Infusions. Tea bags were infused for varying lengths of time (1, 3, or 5 minutes) and the infusions obtained were analyzed for the amount of solids extracted and their organoleptic qualities. The results (Table 5, part A) suggest that tea aroma is extracted faster than the astringent principles (i.e. the polyphenols), and that the tangy portion of the astringency is not extracted as fast as the non-tangy portion of the astringency. However, the overall "tea taste" of the infusions appears to be determined by a combination of the aroma and the astringent principles. These indications were verified in the following further experiments:-

The Effect Of Tea Aroma On The Taste Of Black Tea Infusions. First, the aroma was removed from the tea infusions (Table 5, part A) by stripping off the volatile materials present in the tea infusions under reduced pressure. Removal of the aroma from the tea infusions was found (Table 5, part B) to reduce the overall tea-like quality of the tea infusions, but it had virtually no effect on the level of astringency in the infusions. The effect of removing aroma from the tea infusion could be reversed by adding the aroma back to the stripped infusions.

The Effect Of Degallating The Tea Polyphenols On The Taste Of Black Tea Infusions. The importance of galloyl groups (VIIb) on the tea polyphenols in determining the amount of astringency of these polyphenols was clearly indicated by results obtained by tasting individual tea polyphenols (see discussion above of results summarized in Table 3). The importance of galloyl groups on tea polyphenols to the taste of whole black tea infusions was tested by treating the whole black tea infusion with a purified preparation of the enzyme tannase (EC. 3.1.1.20). This enzyme is an esterase that acts specifically on the ester bond between galloyl groups (VIIb) and gallated tea polyphenols (II, IV, VIII, IX, XII-XIV, XIX, XX, and others) (29, 30, 31).

Degallating a 3-minute infusion of black tea leaf (Table 5, part C) completely eliminated the tangy portion of the astringency of the infusion but had no effect on the non-tangy portion of the astringency. Dearomatizing the degallated tea infusion had the

Table 5:   The Effect of Various Treatments on the Taste of

Description of the Tea Infusion [a]	Total Amount of Tea Solids in Infusion (mg)
A. Effect of length of infusion period:	
Tea leaf infusion - 1 minute	490
" " " - 3 minutes	590
" " " - 5 minutes	670
B. Effect of aroma removal (stripping):	
Tea leaf infusion - 3 minutes	590
Stripped infusion (aroma removed)	590
Aroma	0
Reconstituted stripped infusion + aroma	590
C. Effect of degallating tea polyphenols (i.e., treating	
Tea leaf infusion - 3 minutes	590
Infusion after degallating	590
Stripped infusion after degallating	590
Degallated stripped infusion + aroma	590
D. Effect of decaffeination (i.e.,black tea leaf decaffein-	
Infusion (3 minutes) of decaffeinated (and	
dearomatized) tea leaf	535
Decaffeinated infusion + caffeine	580
Decaffeinated infusion + aroma	535
Decaffeinated infusion + caffeine + aroma	580
E. Effect of milk (i.e., adding 1 teaspoon milk to tea infu-	
1 minute infusion + milk	490
3 minutes infusion + milk	590
5 minutes infusion + milk	670
F. Effect of adding lemon juice (i.e.,juice squeezed from a	
Tea leaf infusion - 3 minutes (pH 4.8)	590
Infusion + 3 ml lemon juice (pH 3.2)	590
Infusion + HCl (pH 3.2)	590

[a]   2.27g black tea leaf was infused in a tea cup with 6 oz.
[b] Key to ratings: 0 = none; 1 = threshold; 2 = weak; 3 =

Table 5 (cont.)

Black Tea Infusions

Amount of Caffeine in Infusion (mg)	Organoleptic Evaluation [b]			
	Astringency		Aroma	Black Tea Taste
	Tangy	Non-Tangy		
41	1	2	3	Mild
48	3	3	3	Strong
54	4	4	4	Very strong
48	3	3	3	Strong
48	3	2	1	Flat
0	0	1	2	Flavory, pungent
48	3	3	3	Strong
the tea infusion with tannase enzyme):				
48	3	3	3	Strong
48	0	3	3	Mild
48	0	2	1	Weak
48	0	3	3	Mild
ated, and consequently dearomatized, by solvent extraction):				
3	1	2	1	Weak
48	3	2	1	Mild
3	1	3	3	Mild
48	3	3	3	Strong
sions from A above):				
41	0	2	3	Weak, milky
48	0	3	3	Mild, milky
54	1	3	3	Strong, milky
lemon wedge):				
48	3	3	3	Strong
48	1	2	3	Mild, lemon
48	1	2	3	Mild, sour

freshly boiled tap water producing on average 5.2 oz. of beverage.
moderate; 4 = strong.

effect of reducing both the tangy and the non-tangy portions of
the astringency and reducing the overall tea-like quality of the
infusion to a barely perceptible level. The effect of dearomatizing the infusions was entirely reversed by restoration of the
aroma removed.

The Effect Of Caffeine (XXI) On The Taste Of Black Tea Infusions. Samples of decaffeinated black tea leaf and regular
black tea leaf were brewed side by side and the infusions obtained
were chemically analyzed and organoleptically evaluated. The
results (Table 5, part D) indicated that removal of caffeine from
the tea infusion has a significant effect on the taste of the infusion. Specifically, decaffeination causes the bitterness of a
black tea infusion slightly to increase and decaffeination changes
the nature of the astringency in the infusion from the tangy type,
which is characteristic of black tea, to a non-tangy type. Further, the results (Table 2, part D) show that degallation of the
tea polyphenols has only a relatively small effect on the taste
of decaffeinated tea beverages (this effect is a small general
decrease in all taste properties), whereas degallation of the tea
polyphenols in a regular tea infusion causes a marked reduction
in the astringency of the infusion.
      It has long been known that caffeine (the predominant xan-
thine compound in tea; see Table 1) complexes with tea polyphenols.
In fact, the complexation of black tea polyphenols and caffeine is
responsible for much, but not all, of the tea cream formation
(i.e. the precipitation of tea solids) that occurs when black tea
infusions cool down ($\underline{32}$, $\underline{33}$, $\underline{34}$). A more detailed investigation
by Collier et al. ($\underline{35}$) showed that condensation of the tea flava-
nols (ex. I + III + $O_2 \rightarrow$ X) and the presence of galloyl groups
(VIIb) on the tea polyphenols, decreases the solubility of caf-
feine/tea polyphenol complexes.
      Our results (Table 5, part D) suggest that caffeine com-
plexes with the black tea polyphenols in a way that prevents
these polyphenols from complexing with themselves to form larger
polyphenolic molecules. These caffeine/black tea polyphenol complexes are less soluble in cold water than the internal black tea
polyphenol/polyphenol complexes, and the caffeine/polyphenol complexes have a more sharp tangy astringency than the polyphenol/
polyphenol complexes which have a lingering mouth-drying, mouth-
coating effect (i.e. non-tangy astringency).
      The ability partially to transform the astringency in regular
black tea infusions (containing caffeine) to something very close
to the astringency of the decaffeinated tea infusions by degallating the tea polyphenols (Table 5, parts C and D) suggests that
the galloyl groups on the black tea polyphenols are the specific
sites involved in the complexation with caffeine, or other black
tea polyphenols. If the above is true, then it is also true that
the galloyl groups on the tea polyphenols are critical determinants of the type of astringency that exists in tea beverages. It

is noteworthy that it is now known (36, 37, 38) that degallation
of black tea polyphenols will eliminate most of the black tea
cream furnishing another piece of evidence for the importance of
the specific nature of the complexation between caffeine (XXI)
and the galloyl groups (VIIb) of black tea polyphenols (II, IV,
VIII, IX, XII-XIV, XVI, XIX, XX, and others).

The Effect Of Milk On The Taste Of Tea Infusions.   Milk is
often added to black tea infusions to ameliorate the taste of the
beverage.   The effect of this practice was studied by adding milk
to black tea infusions and determining the organoleptic properties
of the resulting beverage.   The results (Table 5, part E) showed
that the addition of milk to black tea infusions caused a marked
lowering of the astringency of the infusions.   The reduction of
astringency in these experiments was complete with the 1-min.
and 3-min. infusions and almost complete with the 5-min. infusion.
Of course, the effect noted here will be highly dependent on the
amount of tea solids in the cup and the amount of milk added.
    It is noteworthy that the addition of milk to these tea in-
fusions had practically no effect on either the overall tea-like
quality or the aroma.   The addition of milk to black tea infusions did cause other effects on the taste of the tea infusions
(not noted in Table 5); such as an increase in the body of the
beverage, contribution of a smoothness to the taste, and contribution of a milky taste per se; but these effects are foreign to
the subject of this discussion.
    Since the astringency of black tea infusions is established
to be due to the phenolic compounds present, it can be safely as-
sumed that the milk has its effect by tying up the tea polyphenols
in such a way that they no longer have astringent properties: The
milk proteins are prime candidates for the agents in milk that
cause this tying up of the tea polyphenols.   Brown and Wright (39)
isolated milk protein/black tea polyphenol complexes and studied
their electrophloreic properties.   The interaction of proteins
with polyphenolic compounds is a well known, frequently observed
phenomena (40), but it is important to this discussion to note
two consequences of the reaction between tea polyphenols and milk
proteins: First, the complexes do not precipitate as do virtually
all other protein/tea polyphenol complexes.   Presumably, the milk
protein-tea polyphenol complexes form stable colloidal suspensions.
Second, these complexes have the effect of appreciably reducing
the astringency of black tea infusions which is desirable to most
consumers (G.W. Sanderson, unpublished data).

The Effect Of Lemon Juice On The Taste Of Black Tea Infusions.
Lemon juice added to a black tea infusion was found (Table 5,
part F) to cause a marked reduction in the astringency of the
beverage.   Further, the tangy part of the tea astringency was
more affected than the non-tangy part.   The overall effect of
this reduction in astringency was to reduce the tea taste im-

pression of the beverage from "strong" to "mild".

The pH of a second black tea infusion was adjusted down with hydrochloric acid from an initial pH 4.8 to pH 3.2 (the same pH change effected by addition of the lemon juice described above). This tea beverage with pH adjusted by means of an inorganic acid was found (Table 5, part F) to have virtually the same taste properties as the tea beverage with lemon juice, especially as regards the astringency and the strength of the tea taste.

These results indicate that slightly reducing the pH of a tea beverage (such as from pH 4.8 to about 3.2) will reduce the astringency of the beverage and that this change is most pleasantly accomplished by adding a little lemon juice: This, of course, has the added advantage of contributing a touch of lemon flavor which complements the tea flavor. The fact that the tangy part of the tea astringency is affected most suggests that it is the caffeine/polyphenol complex that is most altered by the pH change. The chemistry underlying this phenomena is not understood, but the phenomena, i.e. the reduction in astringency caused by lowering pH, has been described more than once before (40).

Summary And Conclusions

The results of our investigations confirm and extend earlier research (9, 15, 41, 42, 43) that indicates the prime importance of the tea polyphenols in determining the taste of black tea infusions (beverages). Through the process of tea fermentation (Figures 1 and 2) the green tea flavanols (I-VI) are oxidized and condense to form the theaflavins (XI-XIV), the thearubigins (VII-XX, and other unknowns), and other minor products (VIII-X, XV, XVI, and other unknowns). These changes are accompanied by changes in the taste of infusions from green, grassy, harsh, bitter, with slight non-tangy astringency (fresh green tea flush), to green, very harsh, with strong non-tangy astringency (fermented but not fired), to flowery, slightly green, with pleasant tangy astringency, and mild black tea flavor (fermented and fired). These changes were found to be definitely associated with the oxidation of the tea flavanols in that elimination of both the tea fermentation and the firing processes prevented the development of a characteristic black tea taste (Table 3) and removal of the polyphenols from a black tea infusion effectively removed all recognizable black tea character (Table 2).

Firing of the fermented tea flush material was shown in our investigation (Table 3) to be essential to the development of black tea flavor. This finding has been reported previously by Bokuchava et al. (17) and by Bhatia and Ullah (18). The role of firing in the development of black tea flavor is not well understood but the available evidence suggests that the following changes are brought about by firing that are important in this context: (a) Firing following tea fermentation causes some further changes in the polyphenols that resemble the changes taking place during tea fermentation, i.e. changes in the amounts of theaflavins and thearubigins. However, these changes which take place at higher temperatures and in a more concentrated environment may be qualitatively different from those occuring during tea fermentation itself. (b) Firing causes considerable insolubilization of tea solids and the materials insolubilized include both polyphenols and non-polyphenolic compounds. The polyphenolic materials lost to the infusion in this way are mostly thearubigins and the non-polyphenolic materials lost are probably small amounts of peptides (44) and polysaccharides (45). These losses may be most important especially if it were found that the compounds with the harshest, strongest non-tangy astringency were perferentially lost in this process. (c). The formation of black tea aroma is entirely dependent on tea fermentation and firing (16, 17, 18, Table 3), and black tea aroma was found to be an essential complement to the black tea solids for the expression of full black tea flavor (Table 5). It is also known that firing drives off appreciable amounts of aroma constituents (17, 46), and this may lead to an improved balance of aroma constituents as far as black tea aroma is concerned. The above 3 points certainly deserve further investigation.

The relationship of galloyl groups (VIIb) and caffeine (XXI) to the tangy astringency of tea infusions is most important (Tables 2 and 5). Tangy astringency is possibly what some other researchers (9, 25, 43), and the tea trade (47) call briskness. In any case, tangy astringency is difficult to define, a fact recognized long ago by Bate-Smith (48) in his review of astringency in food products, yet it is a most important characteristic part of the taste of black tea infusions. Roberts (9, 49) had found that "briskness" in black tea infusions was correlated to some extent with the theaflavins and the caffeine content of these infusions and Wood and Roberts (25) provided some additional evidence in support of this contention. However, we can now say that it is the galloyl groups on the theaflavin gallates (XII-XIV) and other gallated black tea polyphenols (VIII, IX, XVI, XX, and other gallated unknowns) that react with caffeine (XXI) to produce the tangy astringency associated with "briskness" in black tea infusions.

It is noteworthy that studies of consumer practices and preferences in the United States (G.W. Sanderson, unpublished) indicate that tea bags are usually brewed for only about 1 min. Further, the criticism of tea beverages that is obtained more often than any other is that the beverage is too bitter (consumers appear to confuse bitterness with astringency in the case of tea beverages). Apparently, consumers in the United States control the level of astringency in their cup of tea by using a rather short extraction time, thereby limiting the amount of tea solids extracted. As shown in Table 5, part A, this is an effective means of minimizing the astringency of the infusion while

at the same time providing for a reasonable amount of tea flavor
to be extracted. Of course, reducing the pH of the black tea infusion by adding lemon juice is also a sound means of reducing
the astringency of the infusion (Table 5, part F) and this too is
a common practice in the United States.

In many places outside the United States, it is customary to
brew tea much stronger than within the United States (by brewing
for 3 to 5 min. rather than for 1 min. and/or by using more tea
to prepare a serving). However, it is also customary to add milk
to tea infusions in these places. The United Kingdom, India,
and Sri Lanka are good examples of countries where such practices
are almost universal. The results shown in Table 5, part D, indicate that the high level of astringency naturally associated
with the stronger tea infusions preferred in many countries outside the United States is neutralized with milk, rather than be
liked or tolerated, by the consumers in these other countries.

On The Chemistry Of The Taste Of Green Tea

A detailed discussion of green tea is outside the scope of
this paper. However, attention should be drawn to a recent paper
by Nakagawa (50) that provides much useful information on the
chemistry underlying the taste of green tea infusions. Nakagawa's
(50) results indicate that the major components of taste in green
tea are bitterness, astringency, brothy, and sweetness. The
bitterness and astringency was shown to be due to the green tea
polyphenols: The tea flavanols (I-VI), especially the gallated
flavanols (II, IV), and leucoanthocyanins were considered to be
most important in determining these taste characteristics, but
some unidentified phenol-type materials were also thought to make
a significant and desirable contribution to green tea bitterness
and astringency. The brothy taste of green tea was shown to be
due to amino acids, and the sweetness to sugars. Caffeine was
reported to play no significant role in determining the taste of
green tea. The contrast between Nakagawa's (50) results for
green tea and the results discussed above for black tea derive
in part from the different clones of tea plants cultivated for
green tea manufacture, but mostly from the tea fermentation process (Figure 2) which is part of the black tea manufacturing
process but which is purposely prevented in the green tea manufacturing process (51, 52).

Experimental

The beverage strength extract was prepared by brewing black
tea leaf in 75 times its weight of distilled deionized water for
5 min. The extract was freeze-dried from about 2% solution for
use as required.

Aroma Recovery. Tea aroma was recovered by collecting about

25% distillate from a freshly prepared tea beverage. Distillation was carried out in a rotory evaporator under vacuum at 40°C
and with a condenser temperature at -5°C.

Fractionation. The beverage strength tea extract was concentrated to about 2% (w/w) solids prior to fractionation by solvent
extraction. Caffeine and other purines of tea were removed by
extracting the tea concentrate with trichloroethylene (TCE) for
48 hr. using a liquid-liquid extractor. After the extraction was
completed TCE was distilled under vacuum to recover caffeine.
Traces of TCE were removed from the aqueous tea extract under
vacuum using a rotory evaporator. Next, tea polyphenols were
recovered from the caffeine free tea extract by ethyl acetate
(EtAc) extraction for 48 hr. using a liquid-liquid extractor.
Polyphenols were recovered from the EtAc fraction by removing
EtAc under vacuum after some distilled water was added to that
fraction. Added water was removed from the polyphenol fraction
by freeze-drying. The caffeine and polyphenol free aqueous
fraction of the tea was then freeze-dried after removal of traces
of EtAc under vacuum.

A tea extract free of all tea polyphenols was obtained by
passing a freshly brewed tea solution through a column packed
with hydrated polyamide CC6 (Brinkmann). The column was then
washed 3 times with hot distilled water. The eluates obtained
were completely free of polyphenols as judged by paper chromatography. A total polyphenol and caffeine free tea extract was
obtained by passing the above polyphenol free extract through a
presolvated XAD-2 column (53). Purified tea flavanols were obtained or prepared as described previously (54).

Analytical Methods. Organoleptic evaluations were done
using a panel consisting of laboratory personnel. The panel was
trained for this work and all the testing was carried out under
standard conditions.

Minerals were determined by atomic absorption and flame
emission spectroscopy.

Pectins were determined by the method of McComb and
McCready (55).

Sugars were determined by a modification (R. Simons, unpublished) of a procedure by which sugars are purified by ion
exchange chromatography (56) and determined by quantitative gas
chromatography (57).

Organic acids were separated by a modification (R. Simons,
unpublished) of a procedure by Fujimaki et al. (58), and determined by gas chromatography (59).

Amino acids were determined by automatic amino acid analyzer.

Caffeine was determined by g.l.c. after chloroform extraction (P.D. Collier, unpublished method).

Individual flavanols were estimated using the method of
Collier and Mallows (60).

Theaflavin and thearubigins analysis of tea solutions was

carried out using the method of Roberts and Smith (15).
    All other analytical methods were official A.O.A.C. methods
(61).

    Black Tea Manufacture: Freshly harvested green tea flush
was air-freighted so as to reach our laboratory in the evening
of the day of harvesting (54). Black tea manufacture (62, 63)
was carried out according to the following laboratory scale procedure :
    The flush was spread out on a bench top to wither overnight
to a moisture content of about 65%. The withered flush was
macerated by passing it 3 times through a roll mill. The macerated
tea flush was spread about 3 cm. deep in trays covered with damp
cheesecloth and allowed to undergo tea fermentation for the
times specified. At the end of the desired tea fermentation
period, one sample of fermented tea flush was frozen by mixing
with crushed dry ice and freeze dried, and one sample was fired
by forcing air at 97°C through the sample for about 25 min. All
samples were dried to a moisture content of about 5% which ren-
dered them stable and ready for chemical analysis and organoleptic evaluation.

Literature Cited

1. Hainsworth, E. "Encyclopedia of Chemical Technology," 2nd
    ed., Standen, A., Ed., pp. 743-755, Wiley (Interscience),
    New York, 1969.
2. Sanderson, G.W. "Structural and Functional Aspects of Phy-
    tochemistry," Runeckles, V.C., Tso, T.C., Ed., pp. 247-316
    Academic Press, New York, N.Y., 1972.
3. Peterson, M.S. "Encyclopedia of Food Technology," Johnson,
    A.H., Peterson, M.S., ed., pp. 889-891, AVI Publishing Com-
    pany, Westport, Connecticut, 1974.
4. Sreerangachar, H.B. Biochem J. (1943) 37, 661.
5. Kursanov, A.L. Kulturpflanze Beiheft (1956) 1, 29.
6. Vuataz, L., Brandenberger, H. J. Chromatog. (1961) 5, 17.
7. Millin, D.J., Rustidge, D.W. Process Biochem. (1967) 2, 9.
8. Burg, A.W., Tea and Coffee Trade Journal (1975) 147, January,
    40.
9. Roberts, E.A.H. "Chemistry of Flavonoid Compounds," Geissman,
    T.A., Ed., pp. 468-512, Pergamon Press, London, 1962.
10. Berkowitz, J.E., Coggon, P., Sanderson, G.W. Phytochem. (1971)
    10, 2271.
11. Sanderson, G.W., Berkowitz, J.E., Co, H., Graham, N.H.
    J. Food Sci. (1972) 37, 399.
12. Brown, A.G., Eyton, W.B., Holmes, A., Ollis, W.D. Nature
    (London) (1969a) 221, 742.
13. Brown, A.G., Eyton, W.B., Holmes, A., Ollis, W.D. Phytochem.
    (1969b) 8, 2333.
14. Weinges, K., Muller, O. Chemiker Zeitung (1972) 96, 612.

15. Roberts, E.A.H., Smith, R.F. J. Sci. Food Agric. (1963) 14, 689.
16. Saijo, R., Kuwabara, Y. Agr. Biol. Chem. (1967) 31, 389.
17. Bokuchava, M.A., Skobeleva, N.K., Dmitriev, A.F. Dokl. Akad. Nauk SSSR (1958) 115, 183.
18. Bhatia, I.S., Ullah, M.R. J. Sci. Food Agric. (1965) 16, 408.
19. Sanderson, G.W., Graham, H.N. J. Agric. Food Chem. (1973) 21, 576.
20. Sanderson, G.W. "International Symposium: Odour and Flavour Substances," Drawert, W., Ed., pp 65-97, Haarmen and Reimer, GmbH, Holzminden, W. Germany, 1975.
21. Suzuki, T., Takahashi, T. Biochem. J. (1975a) 146, 79.
22. Suzuki, T., Takahashi, T. Biochem. J. (1975b) 146, 87.
23. Roberts, E.A.H., Cartwright, R.A., Oldschool, M. J. Sci. Food Agric. (1957) 8, 72.
24. Coggon, P., Moss, G.A., Sanderson, G.W. Phytochem. (1973) 12, 1947.
25. Wood, D.J., Roberts, E.A.H. J. Sci. Food Agric. (1964) 15, 19.
26. Biswas, A.K., Biswas, A.K., Sarkar, A.R. J. Sci. Food Agric. (1971) 22, 196.
27. Biswas, A.K., Sarkar, A.R., Biswas, A.K. J. Sci. Food Agric. (1973) 24, 1457.
28. Hilton, P.J., Ellis, R.T. J. Sci. Food Agric. (1972) 23, 227.
29. Dykerhoff, H., Armbruster, R. Hoppe Seyler's Zeitschrift fur Physiologische Chemie (1933) 219, 38.
30. Haslam, E., Strangroom, J.E. Biochem. J. (1966) 99, 28.
31. Iibuchi, S., Minoda, Y., Yamada, K. Agr. Biol. Chem. (1972) 36, 1553.
32. Roberts, E.A.H. J. Sci. Food Agric. (1963) 14, 700.
33. Wickremasinghe, R.L., Perera, K.P.W.C. Tea Quart. (1966) 37, 131.
34. Smith, R.F. J. Sci. Food Agric. (1968) 19, 530.
35. Collier, P.D., Mallows, R., Thomas, P.E. Phytochem. (1972) 11, 867.
36. Tenco Brooke Bond Ltd. British Patent 1,249,932 (1971).
37. Takino, Y. New Zealand Patent 160,729 (1971).
38. Takino, Y. Canadian Patent 905205 (1972).
39. Brown, P.J., Wright, W.B. J. Chromatog.(1963) 11, 504.
40. Joslyn, M.A., Goldstein, J.L. Advan. Food Res. (1964) 13, 179.
41. Bokuchava, M.A., Novozhilov, N.P. Biokhim. Chainogo Proiz-vodstva, Akad. Nauk SSSR (1946) 5, 190.
42. Bokuchava, M.A., Skobeleva, N.I. Advan. Food Res. (1969) 17, 215.
43. Millin, D.J., Crispin, D.J., Swaine, D. J. Agr. Food Chem. (1969) 17, 717.

44. Wood, D.J., Bhatia, I.S., Chakraborty, S., Chondhury, M.N.D., Deb, S.B., Roberts, E.A.H., Ullah, M.R. J. Sci. Food Agric. (1964) 15, 14.
45. Sanderson, G.W., Perera, B.P.M. Tea Quart. (1965) 36, 6.
46. Yamanishi, T., Kobayashi, A., Sato, H., Nakamura, H., Uchida, A., Mori, S., Saijo, R. Agr. Biol. Chem. (1966) 30, 784.
47. Palmer, D.H. J. Sci. Food Agric. (1974) 25, 153.
48. Bate-Smith, E.D. Food (1954) April, 124.
49. Roberts, E.A.H. J. Sci. Food Agric. (1958) 9, 381.
50. Nakagawa, M. Nippon Shokuhin Kogyo Gakkai-Shi (1975) 22, 59.
51. Ukers, W.H. "All About Tea", 2 vol., Tea and Coffee Trade Journal Co., New York, N.Y., 1935.
52. Ukers, W.H., Prescott, S.C. "Chemistry and Technology of Food and Food Products", 2nd ed., Vol. 2, Jacobs, M.B., Ed., pp 1656-1705, Interscience Publ., New York, N.Y., 1951.
53. Gustafson, R.L., Albright, R.L., Meisler, J., Lirio, J.A., Reid, O.T., Jr. Ind. Eng. Chem. Prod. Res. Develop. (1968) 7, 107.
54. Co, H., Sanderson, G.W. J. Food Sci. (1970) 35, 160.
55. McComb, E.A., McCready, R.M. Anal. Chem. (1952) 24, 1630.
56. Cartwright, R.A., Roberts, E.A.H. J. Sci. Food Agric. (1954) 5, 600.
57. Larson, P.A., Hobbs, W.E., Honold, G.R. Instrum. Food Beverage Ind. (1972) January, 1.
58. Fujimaki, M., Kim, K., Kurata, T. Agric. Biol. Chem. (1974) 38, 45.
59. Fernandez-Florez, E., Johnson, A.R., Fitelson, J. J. Assoc. Off. Anal. Chemists (1970) 53, 1193.
60. Collier, P.D., Mallows, R. J. Chromatog (1971) 57, 29.
61. Horwitz, W. "Official Methods of Analysis," 12th ed., Assoc. Off. Anal. Chemists, Washington, D.C., 1975.
62. Eden, T. "Tea", 2nd Ed., Longmans, Green and Co., Ltd., London, England, 1965.
63. Harler, C.R. "Tea Manufacture", Oxford Univ. Press, London, England, 1963.

Wine Flavor and Phenolic Substances

V. L. SINGLETON and A. C. NOBLE

Department of Viticulture and Enology, University of California, Davis, Calif. 95616

Phenolic substances are very important to the taste and odor of wines. They make direct contributions to flavor which may be very large and overriding as in some astringent red wines, but can be very subdued compared to the effects of other compounds in wines of other types. In dry red wines they usually are the most plentiful constituents after alcohol, tartaric acid, and unfermentable sugars. Indirect effects of phenols on flavor can be large and both psychological and chemical. For example, sensory quality of wine is directly influenced to a considerable degree by color and the influence of color on judgement of flavor is difficult to avoid. The colors of wines are largely the result of anthocyanins, other phenols, and their reaction products whether the wines are pink, red, purplish, red-orange, yellow, golden, or amber. Reactions associated with the oxidation of the wine's phenols normally produce taste and odor changes as well as color changes. Therefore, wine that appears oxidized from its color is likely to be judged as oxidized in flavor even if it is not. In most wines, polyphenols are the main pool of substances capable of autoxidation under normal wine-aging conditions (ambient or lower temperature, about pH 3.3, air access often restricted and at ambient pressure, etc.). Thus, processing and aging of wine produce direct taste or odor changes by modifying the phenols themselves, but also produce indirect flavor effects through associated reactions.

The ideal situation has not been reached wherein we know the complete specific qualitative and quantitative phenolic composition of wines and these substance's individual and collective role in flavor. Considerable progress is currently being made to this goal. The phenolic substances of wine and their direct and indirect effects in wine have been reviewed (1). Further effects associated with coloration and dis-coloration reactions have also been reviewed (2). In this paper the focus is on the specific, direct effects of wine

phenols on odor and taste of wine emphasizing the present
status of our knowledge with only a few observations on the
reactions of phenols of indirect but great importance to wine
flavor. Color as such is ignored. In an effort to be complete and to encourage further work, we have included pertinent
findings not directly involving wines and have deliberately
included some observations and opinions not solidly based on
proven fact. Data drawn from the literature are documented,
but not exhaustively. Facts considered common knowledge in
enology can be checked in general texts (e.g., 3).

I. Phenols as Flavorants. What effects on wine taste and
odor might be expected from the general knowledge of phenols
and flavor? Certain phenols are known to have several different types of direct contributions to flavor: odorants,
pungents, sweet substances, bitters, and astringents.
Odorants must, of course, be reasonably volatile. To interact
at the odor sensing site and cause an odor sensation, appreciable lipid solubility and some water solubility are
considered requisites (4). Many small or simple phenols
can be seen to fill these requirements well. Odorous phenols
of some interest in foods range from phenol, cresols, and
guaiacol with "medicinal," "phenolic," or possibly smoky
odors to more characteristically pleasant odorants such as
vanillin and methyl salicylate. Flavonoids and other sizeable
phenols have no significant odor in the pure state nor do
highly polar phenol derivatives, glucosides or gallic acid
for example, which have high water solubility or low vapor
pressure. In wine, the odorous phenols would be sought in
the volatile, solvent extractable, nonflavonoid fraction.
        Pungency is considered a "hot," penetrating, burning
sensation in the mouth which at lower levels may be "warm,"
spicy, sharp or harsh. It is also an odor descriptor and part
of the burning, penetrating, harsh, spicy "phenolic" odors of
some phenols is believed another expression of pungency in
volatile compounds. Pungency is not confined to phenols;
consider, for example, acrolein, cinnamaldehyde and even high
concentrations of ethanol which give burning "taste" sensations. Many phenols have a pungency component in their
flavor and notable pungents include eugenol from cloves,
gingerols from ginger, and capsaicin from chile peppers.
These latter compounds and some pungent synthetic analogs all
are ortho-methoxyphenols with a nonpolar para side chain.
Pungent phenols are also expected to be in the nonflavonoid
fraction extractable with immiscible solvents such as ethyl
acetate, but may (eugenol) or may not (gingerols) be volatile
even with steam (5).
        Astringency is a contracting (puckering), drying, mouth
feeling involving precipitation of the proteins of saliva and
the mucous surfaces (5, 6). Astringency among food constit-

uents is almost entirely from "tannins," the larger natural polyphenols. Based partly on their ability to precipitate proteins, polyhydroxyphenols with a molecular weight of the order of 500 are considered the minimum for appreciable astringency and astringency generally increases with increased degree of polymerization up to some point or the solubility becomes limiting (1, 6).

None of the effects mentioned have involved the primary tastes: salt, sour, sweet, and bitter. The salty taste is rarely of any significance in wine, but the sour taste is quite important. Substances giving salty or sour tastes are ionizable (7) and phenols without carboxy groups (which do not ionize appreciably under wine conditions) are not directly involved in these tastes. Sweetness can be and bitterness often is a property of phenolic substances. A few simpler phenols like phloroglucinol are somewhat sweet, but ordinarily the sweetness is weak compared to sugars and accompanied by "medicinal" or other flavors. A few phenols derived from natural flavonoids are intensely sweet, naringin dihydrochalcone, for example. There is, however, no evidence of a significant sweetness contribution by any phenol found in grapes or wine.

It is not uncommon that sweet, bitter, and tasteless members occur within the same series of compounds and this is true of phenols. Naringin is the intensely bitter 7-β-neohesperidoside of naringenin; 4',9,7-trihydroxyflavanone. It is the predominant bitter substance in grapefruit and can be converted to the intensely sweet dihydrochalcone. Naringenin 7-β-rutinoside is tasteless (8, 9). Current theories indicate that sweetness requires two electronegative atoms such as oxygen about 3 angstroms apart, one with an available proton and the other as a potential proton acceptor for intermolecular hydrogen bonding with the receptor site (7). Bitter substances appear to have a similar arrangement except that the proton to negative center distance is reduced about half, producing intramolecular hydrogen bonding and relative hydrophobicity (7).

The spacial arrangement of the key groups is also important in whether a given compound will be sweet or bitter and how intensely so. It seems pertinent that in the inositol series substances with more than 4 hydroxy groups are only sweet whereas those with less than 4 were bitter and one with 4 hydroxyls was tasteless (10). The picture of a water soluble hydroxylated ring with some hydrophobic character fits natural phenols well and many phenols have some bitter character if they have flavor at all. A few examples include picric acid, matairesinol glucoside in safflower (11), an isocoumarin in carrots, oleuropein in olives, and 2,4,5-trimethoxybenzaldehyde in carrot seed (12). Of course, many terpene, tropolone, glycoside, alkaloid, and peptide derivatives are also bitter,

but no bitters other than phenolic have, to our knowledge,
been demonstrated in wines unless they were deliberately
flavored as with vermouths.

II. Sources of Phenols in Wines. The major source of phenols
in wine is, of course, the grapes used to make the wine. The
natural phenols preexisting in the grape may be modified by
the enzymes and exposure incident to crushing and preparation
for fermentation. The yeasts increase phenol content by conversion of nonphenolic substrates to phenolic derivatives and
will modify the phenolic mixture by influencing extraction
or solubility through the alcohol produced, by adsorption
and precipitation with the yeast cell, and by metabolism
into new phenols. Other microorganisms, particularly the
malic acid fermenting lactic acid bacteria and perhaps
organisms causing incipient spoilage, may also generate or
modify wine's phenols.

Processing, particularly operations such as oxidation
(as in sherry making) or pasteurization, modifies phenols in
wine. Aging produces qualitative and quantitative changes
in the phenolic makeup of wine and particularly fine wines
are often aged for several years. The first stage of aging
is often in wooden containers. Depending on prior container
useage, surface of wood per unit of wine, and time of contact,
additional phenols are contributed from the container wood to
the wine (13b). Finally, phenols might be in wine as inad-
vertant contaminants or as part of flavors added to such wines
as vermouths.

III. Total Phenol Contents of Wines. The total concentration
of all phenols can be determined quite satisfactorily in substances such as wine by molybdotungstophosphoric colorimetry
(14). The results are expressed as mg of gallic acid equivalent
(GAE) per liter of wine. By comparison with suitable known
phenols the contribution of any other phenol to the total mg
GAE/l of the wine can be predicted satisfactorily (13a). For
example, (+)-catechin behaves as an equimolar mixture of
phloroglucinol and catechol and gives 1.5 times the molar
color yield of gallic acid in this assay.

If the phenols of grape berries are exhaustively extracted
with aqueous ethanol, the total is considerably dependent upon
the grape variety as shown in table 1 (15).

Table I
Total Extractable Phenols in Ripe Grape Berries

Berry color	Variety	°Brix	Total phenol mg. GAE/kg. fresh wt.
White	Aligoté	20.7	3240
"	Emerald Riesling	20.6	5470
"	French Colombard	18.7	2530
"	Muscat Alexandria	18.8	2840
"	Pinot blanc	20.1	6060
"	Sauvignon blanc	21.8	2090
"	Sémillon	19.8	2430
Red	Calzin	21.0	4980
"	Catawba	22.1	4060
"	Grenache	20.0	2850
"	Petite Sirah	20.7	4250
"	Delaware	20.6	4360

It follows from these figures that if all the phenols present in the grape berry were freely extracted into the wine the phenol content would be about 2000–6000 mg/l. These values may be somewhat low since grapes grown in cooler regions have more total phenol and these grapes were from a warm area. Wines with 6500 mg GAE/l have been prepared when we deliberately attempted to raise the total phenol as much as possible. Such wine is undrinkably astringent. If cluster stems were extracted, additional total phenol content could be contributed up to about 2000 mg GAE/l. Stems give to wine a somewhat hot or peppery character the chemistry of which has not been clarified (1).

The data available from several sources (1) suggest about 5500 mg GAE/kg total phenol in the average red wine grape and about 4000 mg GAE/kg in the average white wine grape. The difference is partly that the white grapes do not make anthocyanins and this portion of the total phenol appears to be lost rather than diverted to other phenolic derivatives. Furthermore, the red berries tend to be smaller, thus have more seeds and skins for a given weight of fruit (16). The total phenol was distributed about 3.3% in the skins, 0.7% in the pressed flesh, 3.4% in the juice and 62.6% in the seeds for a series of red grapes and 23.2, 0.9, 4.5, and 71.4% respectively in a series of white wine grapes (1). Seedless grapes' total is reduced about 2/3 by the absence of the seeds.

Initial winemaking practices greatly influence the amount of phenols which transfer from the grape to the wine. Light white wines of high quality are usually made by separating the juice immediately upon crushing the grapes. The contact with pomace (skins and seeds) is thus minimal and

the pressing is restrained in juice recovery. The juice is
frequently clarified prior to fermentation. The phenols of
such wines are those of the juice with minimal transfer from
the skins or seeds. The total phenol content is usually about
50 to 350 mg GAE/1 in these wines, averaging about 250 mg
GAE/1 (1). If the grapes are overripe, shriveled, heated,
frozen or otherwise damaged so as to cause extraction of the
phenols from the solid parts, the total phenol content will
rise. The same is true of prolonged pomace contact before
juice separation or excessive pressing for juice recovery.
Good white table wine may be deliberately made with some
pomace contact to increase flavor and part of the increase
is phenolic. Lower quality white wines may have triple the
usual phenol content. White dessert wines and sherries are
normally made from riper grapes and sometimes with other
conditions such that the total phenol content is frequently
higher, 300 to perhaps 800 mg GAE/1.

Pink wines are ordinarily made by fermenting the mixed
whole mass of destemmed, crushed red grapes for about one day
to extract some anthocyanins from the skins. The time before
fluid separation from the fermenting mixture is extended
for red wines; perhaps 3-4 days is typical at present. Maximum red color release into the wine is reached well before
maximum phenol extraction. This is partly because of slower
diffusion and extraction of molecularly larger tannins
especially from the seeds. About half of the seeds' phenol
content should be extracted in the usual conditions of red
wine making (17). Carignane rosé made by fermentation at
25°C had only 1.65% of the anthocyanin and 7.6% of the berry
total phenols and red wine made by fermenting some of the
same grapes 10 days at 25°C had 26.8% of the berry anthocyanin
and 31.2% of the total phenol (18). If fermentation on the
pomace is continued long enough the total phenol content of
the wine decreases, evidently due to polymerization and precipitation. Bourzeix et al. (19) found the maximum phenolic
content occurred at about 11 days of pomace fermentation.

Considerable data on the typical phenol content of pink
and red wines was reviewed by Singleton and Esau (1). Although
wines from individual grape varieties differ considerably,
rosé wines fermented one day on the skins had about 500 mg/1
of total phenol and about 10% of the red color of wines made
from the same grapes and fermented 3-5 days on the skins to
reach about 1900 mg/1 of total phenol. Selective heating of
grape skins to produce presumably complete release of anthocyanins and other phenols from the skins without contribution
from the seeds produced about 1200 mg GAE/1 total phenol.

The lower content of total phenol in wine compared to
the content in grapes reflects mainly incomplete extraction
from the grape's seeds and skins. However, precipitation,
particularly of the tannins with the grape proteins or yeast

cells also is appreciable in the course of wine making and
clarification. Further precipitation and loss occurs during
aging (1). Berg and Akiyoshi (20) found that a wine prepared
without fermentation by alcohol addition to red juice had
1240 mg/l total phenol. Wine made by fermenting another
portion of the same juice was similar in color but the phenol
content fell to 1020 mg/l. Other studies show even greater
losses of tannin added before fermentation.

There appears to be a sudden transition in total phenol
contents in wines from those under about 500 mg/l to those
above about 1000 mg/l. This seems to represent the buffering
effect of yeast and grape protein tannin precipitation
capacity, followed by a rapid rise to higher levels as this
capacity is exceeded (1). White wines tend to have excess
protein and no true tannin, and red wines have high condensed
tannin content and undetectable protein (21). In one study
on low-color, low-sugar grapes at crushing and at daily
intervals of fermentation on the pomace thereafter the phenol
contents of the samples were 240, 600, 640, 660, 700, 1220
and 1490 mg/l (22). The rise would be faster with more
phenol-rich varieties and also with higher alcohol production
or warmer fermentation temperature.

IV. Total Phenol Content and Wine Flavor and Quality. It
has long been known that increased total phenol content resulting from increased extraction of grape solids during wine
preparation produces increased flavor, particularly increased
astringency in the wine (1). In light white table wines increased phenolic content has usually been associated with
reduced quality of the wine, although stripping of such wines
with efficient phenol adsorbing agents (charcoal, polyamides,
polyvinylpyrrolidinone) is also believed to produce insipid,
lower quality wines. On the other hand, very ripe grapes and
some pomace contact are associated with higher phenol content
and high quality in more robust white table and dessert wines.
Examples of both types are provided by a study of 554 German
wines (23). The average phenol content increased in the order
Mosel, Rheingau, Rheinhessen, Rheinpfalz and Baden and this
is also the series considered to be decreasing in lightness,
elegance, and quality toward heavier wines. However, the
phenol (leucoanthocyanidin) content for wines of increasing
quality and made from increasingly "ripe" grapes averaged
10.6 mg./l. for Kabinett, 12.6 for Spätlese, 17.9 for Auslese,
18.0 for Beerenauslese, and was between 24 and 133 mg./l. for
Trockenbeerenauslese wines, in spite of the fact that Botrytis
cinerea, a mold increasingly involved with grapes for the
latter types, commonly lowers the leucoanthocyanin content
(24). Increased phenol (10 times normal leucoanthocyanidin)
content is also considered responsible for decreased quality
and increased inharmonious, bitter, heavy or coarse taste in
white wines from frosted grapes (25). Over 35 mg/l leuco-

anthocyanidin is rare among German wines except in the late
or botrytized speciality harvests (23). This 35 mg/l would
be included in a total phenol content of about 250 mg GAE/l
(26).

  White wines fermented in the normal fashion from only
juice may not have significantly different quality than those
made with up to 12 hours pomace contact after crushing (27),
but longer contact lowered quality. Phenol content and
astringency rating increase with time of fermentation with
the pomace of white or red grapes as shown by the average
values for 7 sets involving 6 white grape varieties, table
2 (28). Ratings for astringency and bitterness are on a
0-low to 5-high scale and maximum quality rating is 20.

Table 2
Phenol Content and Sensory Ratings, Dry White Wines

| | Pomace Contact During Fermentation | | | |
	0	1 day	2 days	5 days
Total phenol, mg GAE/l	207	261	312	405
Astringency, mean rating	3.4	3.8	3.8	4.5
Bitterness, mean rating	3.3	3.6	3.7	3.5
Quality, mean rating	14.1	11.8	11.7	11.0

  All of the quality ratings (table 2) and all but the 1-
and 2-day astringency ratings were significantly different
(95% confidence), but the bitterness ratings were not. From
these data and comparisons within each individual set of wines
it was concluded that pomace contact sufficient to give about
100 mg GAE/l additional phenol would give a recognizable increase in astringency whereas less would not. That is, the
difference threshold for recognizable astringency increase in
white wine is about 100 mg GAE/l of total grape phenols.
Paired testing rather than sample rating for astringency or
bitterness would probably show a lower threshold. In this
study the highest quality rating for juice-only wine (table 2)
was mainly for nonphenolic reasons, but the association of
high phenol content with decreased quality in dry white wine
agrees with other reports (e.g., 1, 23, 27). Since
astringency increased but bitterness ratings increased only
to a point with increasing phenol, it appears perception of
bitterness was suppressed, presumably by interference by high
astringency (28).

  Red wines cover a much wider range of phenol content
and astringency than whites (1). Astringency is a very significant part of a red wine's character. Mildly astringent
red wines ordinarily have a total phenol content of the order
of 1300 mg GAE/l or less, more robust wines about 1400 mg
GAE/l, and over 2000 mg GAE/l indicates a wine excessively
astringent for most people's preference today. The typical

phenol level in red wines has been moving downward apparently
in response to the public's desire for lighter less astringent
red wine. The trend appears to be world wide at least in wine
we import. About 500 mg/l of total phenol for whites and 2250
mg/l for red California wines were considered preferable in
1935 (1). Commercial California claret had about 2000, 1600,
and 1500 mg GAE/l total phenol in 1935, 1940, and 1946, respectively. California port samples were less than half
these values and burgundy about 20% higher, but showed similar
trends. Wines made experimentally by recommended practices
averaged about 1300 mg/l prior to 1941 and about 1150 mg GAE/l
since 1946. Much of this downward shift was accomplished by
shortening the fermentation time on the skins which probably
has now reached the limit of further reduction for most pro-
ducers. A series of 5 typical and widely sold but low priced
commercial California burgundy wines now on the market averaged
1186 mg GAE/l total phenol. This indicates a very low
astringency level in such wines and the taste effect of the
phenols is lowered still further by the masking presence of
a slight sweetness from residual sugar. Premium dry red table
wines, particularly those intended for appreciable aging, are
still generally 1400 mg GAE/l or more.

On the basis of the data quoted for white wines and considering the typical Weber-Fechner relationships the difference
threshold for astringency in the normal red wine would be expected to be about 250 mg GAE/l of the mixture of phenols
extracted from pomace. This compares rather well with values
estimated by tannin addition (1). Astringency ratings by an
expert panel rose significantly in red wines averaging 1500
mg GAE/l when the flavonoid content increased 280 mg GAE/l,
but differences were not significant when the increase was
only 150 mg GAE/l (16).

V. Nonflavonoids of Wine and Wine Flavor. The content of
phenols which are not flavonoids is relatively constant in
all young wines at about 200 mg GAE/l (29). This group of
compounds accounts for nearly all of the phenols in clarified,
unheated, white grape juice whereas the flavonoids are essentially confined to the skins and seeds. The only appreciable
source of nonflavonoids known from the solid parts of the grape
is the hydroxycinnamic acyl groups which easily hydrolyze from
anthocyanins. Heating or microbial action can also lead to
some conversion of flavonoid to nonflavonoid phenols (1, 30).
Wines aged in wooden containers extract phenols from the wood
that are nearly all nonflavonoid in nature (31).

A. Odorous Phenols of Wine. The content of phenols distillable from wines is normally low. Young white wines had about
1.5-3.2 mg/l calculated as phenol (32). With older wine this
increased to 11.5 mg/l in one 9 years old. Red wines, particularly Georgian style which are stored on the grape pomace

and had 2400-2800 mg GAE/1 total phenol, gave considerably
higher distillable phenol initially 22-36 mg/1, and also increased with aging to 42 mg/1 or so. The small phenols
solvent-extractable without distillation from commercial red
wines were m-cresol, 4-ethylphenol, 4-vinylphenol, 4-ethyl-
guaiacol, and tyrosol (33). Also present in smaller amount
were p-cresol, guaiacol, 4-vinylguaiacol, isoeugenol, vanillin,
and (in one of those wines) 2,6-dimethoxyphenol. Several of
these had been reported previously in wine or wine distillates
along with phenol, salicylic acid and its methyl and ethyl
esters, and vanillic acid (34). All of these plus additional
phenols of similar types have been found in distilled beverages
aged in oak containers or in the oak itself including o-
cresol, 2'-hydroxyacetophenone, 2-ethyl phenol, 4-methyl-
guaiacol, 2'-hydroxy-5'-methylacetophenone, syringaldehyde,
syringic acid, 2-isopropylphenol, 2-ethoxy-4-ethylphenol,
coniferaldehyde, eugenol, ethyl vanillate, and sinapaldehyde
(34).

These compounds may enter wine as extractants from wood,
but some of them evidently can be generated in wine by other
means. Thermal degradation of ferulic acid produces guaiacol
and its 4-methyl, 4-ethyl, and 4-vinyl derivatives, vanillin,
acetovanillone, and vanillic acid (35). Caffeic and p-
coumaric acids would be expected to produce analogous derivatives. These along with a total of 32 phenols are principal
components of smoke and smoky flavors (36). However, these
compounds are produced under conditions much short of pyrolysis
such as mashing of grain for whiskey (37) or heating of apple
juice (38). Certain strains of lactobacilli have been shown
to convert chlorogenic or caffeic and p-coumaric acid in cider
via reduction and decarboxylation to 4-ethylcatechol and 4-
ethylphenol (39). In fact, the same organism can convert
shikimate or quinate to catechol thus generating a new
odorous phenol from nonphenolic precursors (40).

The effect of most of these phenols in grape wine flavor
has not been determined, but information can be drawn from
other studies. In aqueous ethanol solution 4-ethylguaiacol
has a sensory threshold of 0.05 mg/1 with a warm, sweet, spicy,
burnt toffee, phenolic character. Similarly 4-ethylphenol had
a threshold of 1.0 mg/1 and was described as woody, phenolic,
medicinal, and heavy cider in odor (41). These odorants are
considered part of the characteristic odor of bitter-sweet
English ciders since their content is about double the threshold
for 4-ethylphenol and 20-fold that of 4-ethylguaiacol. At high
levels they also contributed to off-flavor. Four-vinylphenol
and 4-vinylguaiacol are reported to have thresholds in water
of 0.02 and 0.01 mg/1 respectively (38). In beer the threshold
of 4-ethylphenol was 0.3 mg/1 (phenolic, astringent); 4-ethyl-
guaiacol was 0.13 mg/1 (phenolic, bitter); and 4-vinylguaiacol
was 0.30 mg/1 (astringent, bitter)(42). The taste thresholds
in water of three major smoke constituents guaiacol, 4-methyl-

guaiacol, and 2,6-dimethoxyphenol were 0.013, 0.065, and
1.65 mg/l and the odor thresholds were up to 60% higher (43).
The combined recognition threshold of the phenol complex
from smoke vapor in water was $\leq$ 6.2 mg/l of phenol by taste
and $\leq$ 21 mg/l by odor while the most desirable concentration
was 27 mg/l by taste or 42 mg/l by odor or less depending on
the method of smoke generation (44).

The sum of the identified volatile phenols extracted by
pentane from Jamaica rum was 1.52 ppm with 4-ethylphenol,
guaiacol and vanillin each at 0.25 ppm (45). The content in
Scotch whiskey of all identified volatile phenols was 0.12
ppm, about 1/3 was eugenol (46). Phenols and cresols give
detectable off flavors when added to beer at 0.03 mg/l (47).

It is obvious that more study is needed directly on
wines, but it seems very probable from these data that the
group of volatile phenols in wines, especially those aged in
oak cooperage, can contribute detectable flavor to wines.
While the content is low and may not reach threshold for any
one component in wine, the flavors of many are similarly
described as spicy, smoky, phenolic, medicinal, etc. and
should be additive in flavor (42). An absorption of only
0.2 mg/l of phenol and cresols into wine from atmospherically
contaminated grapes grown near a factory which emitted these
phenols produced objectionable off-taste in the wine (48).
It also appears that a slight warm, sharp, pungent flavor
may be contributed in part by these compounds. The content
of pungent phenols is much less than in such very hot products as clove oil or ginger oleoresin, but ethanol de-
monstrably potentiates such "hotness."

Vanillin and the related coniferaldehyde, syringaldehyde,
sinapaldehyde, vanillic acid, and ethyl vanillate are believed to be in wine primarily as alcohol extractants and
alcoholysis products from wood (1). In a few very old wines
and brandies vanillin odor becomes recognizable. The recognition threshold of vanillin in water is variously reported
to be 0.5-4 mg/l with the detection threshold about 0.1 mg/l
(49). The threshold of vanillic acid in beer is about 10 mg/l
(50). In brandy aged in American oak barrels the content of
vanillin is about 11 mg/l, syringaldehyde about 16 mg/l, and
all aromatic aldehydes about 55 mg/l calculated as vanillin
(51). Up to 0.25 mg/l of vanillin has been reported in wines
aged in wood, but over 0.5 mg/l was considered as presumptive
evidence of vanillin addition (52). Addition of only 0.05 to
0.5 mg/l of vanillin or p-hydroxybenzaldehyde to synthetic
sake improved the quality (53). It appears and certainly is
the belief of wine tasters that vanillin and related flavors
can be part of the mellowing effect of barrel aging on wine,
but they are not usually present at individually supra-
threshold levels.

B. Flavor Effects of Nonflavonoids of Low Volatility. The odorous, volatile phenols of wine apparently account for only a few ppm of the typical 200 mg GAE/1 of nonflavonoids. The remainder is largely accounted for by cinnamic and benzoic acid derivatives plus some tyrosol and related compounds. Tyrosol is the only phenol known to be produced in significant amounts from nonphenolic precursors by yeast fermentation. It, like other "fusel oil" alcohols, is produced by fermenting yeasts from carbohydrate by decarboxylation and reduction of the α-keto acid analogous to the amino acid, tyrosine in this case (54). It is produced in much larger amounts if tyrosine is present as the major nitrogen source, a condition not found in wine but easily produced artificially. Wine fermentations ordinarily produce more of these higher alcohols when the initial sugar content is higher or the fermentation temperature is warmer. This appears the reason wines usually have a higher tyrosol content than beer. The tyrosol content of wine is occasionally as high as 45 mg/1 (1). Red wines averaged 29 mg/1 and whites 22 mg/1 (55). Age of the wine has no significant effect on the tyrosol content. Tyrosol has an agreeable odor described variously as honey-like, waxy, or old-fruit, but addition of 10-fold the normal content (500 mg/1) did not much accentuate the odor while depressing the flavor quality of white wine (56). Tyrosol is bitter and the taste threshold has been reported to be 10 to 200 ppm in beer (42, 57, 58, 59). Since beer is carbonated and contains considerable background of hop bitters one would expect a lower threshold in white table wine. Tyrosol thus could account for some of the natural bitterness of wine and would account for about 15 mg GAE/1 of the nonflavonoid phenol. Part of the bitterness of apple wine has been attributed to tyrosol (60).

Tyramine occurs in typical wines at 2-3 mg/1, but in a few it has been reported as high as 25 mg/1 (61, 62). It would be expected to have a bitter flavor threshold of about 20 mg/1 based on studies with beer (57).

The remainder, 180 mg GAE/1 or so, of the nonflavonoid in typical wines is largely cinnamic and benzoic acid derivatives. Benzoic acid derivatives account for only a few ppm of the phenol content in white wines. Although they are liberated by alkaline hydrolysis from red wines in the aggregate of up to perhaps 100 mg GAE/1, a major portion of this amount seems to arise from degradation of other phenols and considerably smaller amounts are isolated by simple extraction. Gallic acid is known to occur as flavanol gallates which would liberate free gallic acid into the wine with time (1, 63). The source is unclear of the "bound" forms of the 4-hydroxy-benzoic, protocatechuic, vanillic, syringic, salicylic, and gentisic acids found in traces before but in increased amounts after saponification of wine   (1). Dadic and Belleau (50)

tested all these substances except salicylic acid and found
them to be bitter in water and beer with additional harsh or
other character in some cases. The thresholds in beer were
about 10-50 mg/l for each of these phenols and about half
these values in water. Meilgaard (42) as part of his comprehensive study with purified compounds reported thresholds in
beer of 80 mg/l for vanillic **and** 360 mg/l of gallic acid and
the flavor character was described as astringent. Thresholds
of "astringency" in water are reported to be 30 mg/l for
protocatechuic and vanillic acids, 40 mg/l for gallic, and
240 mg/l for syringic acid (64). Individually, the hydroxy-
benzoic acids appear too low in concentration to produce
appreciable taste contributions to usual wines, but the total
of 30 mg GAE/l or so may well add to the overall bitter impression.

The major phenolic cinnamic acids of wine are caffeic
acid and its derivative esters like chlorogenic acid or caf-
feoyltartaric acid (1). Ferulic acid, p-coumaric acid, and
dihydrocaffeic acid also are present in appreciable amounts
depending on the grape variety or the wine (1, 65, 66).
Estimating from the available data, we assume that the remaining 150 mg GAE/l of nonflavonoid phenol in a typical unaged
wine is about half caffeic acid and its derivatives, and one
quarter each ferulic and p-coumaric derivatives. There are
also small amounts of esculetin, daffnetin and a few other
nonflavonoids which need to be accounted for (1), but which
appear very unlikely to have flavor importance in wine.

The flavor effects of the hydroxycinnamates are evidently mild. Pure chlorogenic acid has only a mild bitter-
sour taste when applied to the tongue. Ferulic acid has a
660 mg/l threshold in beer, p-coumaric had 520 mg/l, and
caffeic acid 690 mg/l in beer with the flavor described as
astringent (42). However, caffeic acid, chlorogenic acid,
ferulic acid and sinapic acid were all reported to have 20
mg/l thresholds in beer in another study (50). The flavor of
ferulic acid was described as sweet in beer but bitter-sweet
in 5% alcohol. The others were bitter or harsh and caffeic
and ferulic were considered astringent. In aqueous solution
the thresholds for "astringency" by the acids were 40 mg/l
for p-coumaric, and 90 mg/l for caffeic and ferulic (64).
Chlorogenic acid at 600 mg/l was rated 3.7 on a 0-5 scale of
increasing bitterness, slightly higher than the same concentration of caffeine (67). In cooked hybrid potatoes where
chlorogenic acid was the major and caffeic acid a minor component, there was a +0.72 correlation for bitterness and
+0.82 for astringency with higher phenol content in the
range 350 to 640 mg/kg fresh weight (68).

Destruction of chlorogenic acid in instant coffee by oxidation is reported to make it milder and better (69) and hydrolysis of chlorogenic acid by tannase is stated to greatly improve the taste of grape juice or wine (70). A pungent taste has been reported to result in synthetic sake from the addition of only 2-3 mg/l of ferulic acid or its ethyl ester (71). Pending additional studies in wine, it appears that there are sufficient amounts of hydroxycinnamates in wines to affect the bitterness and perhaps other aspects of flavor. If the concentration estimate of about 75 mg GAE/l for caffeoyl esters similar to chlorogenic acid is correct, it may be above threshold in wine and considering the probable additive effect of bitterness in the phenolic series the overall influence of these compounds on flavor seems certain.

C. Nonflavonoid Tannins from Wood Aging. The phenols contributed to wine by aging it in oak barrels can be fairly large in amount and include only a small amount of flavonoids (13b, 31). About 1 year in a new 60-gallon European oak barrel can contribute 250 mg GAE/l of nonflavonoids to the wine (31). American oak would contribute about half as much phenol but more oak odor (31, 51). Longer storage will contribute more wood phenols; old brandy may have 2 g/l of extracted solids and 600 mg GAE/l or more phenol. The phenols of oak extract include ellagitannins as major components (72, 73) along with some alcohol soluble lignin fragments and a long list of other phenols in small amounts including esculin, scopolin, their aglycones, umbelliferone and β-methylumbelliferone (73, 74, 75, 76). The conversion of bitter esculin into its less bitter aglycone esculetin is believed one of the reasons stave wood improves with seasoning before making into barrels (76).

The flavor contribution of oak is such that the extract to give a threshold flavor change in a liter of table wine was equivalent to 320-350 mg of oven-dry wood for American oak and 210-270 mg for European oak (13b). This is equivalent to about 22 mg of extracted solids or 7 mg of GAE nonflavonoid phenols for the American oak and 25 mg of extracted solids or 13 mg of GAE nonflavonoids for the European oak (13b, 31). No doubt an important part of the added flavor from the wood extract was from components other than phenols, but it is equally certain that extra flavor contributed by oak phenols can be recognized in oak-aged wines. It has been estimated from the amount of oak extracted in used barrels that about 100-fold the indicated threshold value can be contributed by a barrel to wine (13b) and one year in a new barrel may contribute 20 times the flavor threshold.

VI. Flavonoids and Wine Flavor. The flavonoids in a typical white table wine with a total phenol of 250 mg GAE/l would be of the order of 50 mg/l. Considering 1400 mg GAE/l as the typical total phenol of a red table wine and the nonflavonoid fraction still 200 mg GAE/l, the flavonoid would be about 1200 mg GAE/l. It would be expected that the monomeric flavonoids would diffuse from the grape cells during wine-making more rapidly than the condensed tannins of increasing flavonoid polymerization. There obviously can be some of this effect from the fact already mentioned that the total phenol content increases after anthocyanin color has become maximal in fermentation on the pomace. However, during the 3-4 day pomace contact commonly used today, the extraction of antho-cyanins, flavonols, small and large leucoanthocyanidins and total phenols appear to remain in fairly constant ratios to each other for any given wine (19). For our purposes we can then simplify by assuming that the flavonoids in wine represent a fairly constant mixture of types approximately the same as that in the solid parts of the grape.

The anthocyanins of typical wine grapes total about 500-1000 mg/kg (77) but in young wines they are usually 200-500 mg/l and with aging both break down and combine with tannins into complexes or polymers (1, 2, 78, 79). Anthocyanins when isolated from grapes, purified as a group and placed on the tongue as solid or in aqueous solution as red as wine had only a very mild undistinctive flavor. However, white grapes, made into wines as if they were red, make astringent wines which do not resemble red wines in flavor (28). Since the only appreciable difference appears to be the anthocyanin content of the red wines, their participation in flavor is suggested. This may well be through reactions which occur later rather than directly from the anthocyanin. A purified preparation of cranberry anthocyanins containing 17% antho-cyanins, 18% flavonols and 36% total phenols gave a difference threshold of 290 mg/l when added to cranberry cocktail and produced changes described as increased richness and fruiti-ness (80). We would estimate therefore that unchanged grape anthocyanins in young red wine would have a mild but approx-imately threshold effect on flavor, since they occur in con-centrations exceeding these. A report that young red wine quality correlated positively with increased percent of the anthocyanins ionized rather than anthocyanin total content (81), is interpreted as showing the effect of visible color and pH on wine quality rating rather than a direct effect on flavor by anthocyanins.

The content of flavonols in light white wine is vanish-ingly small and only reaches about 40-310 mg/kg of grapes calculated as rutin (77). Typical content in red wine appears to be about 20-100 mg/l. The ready hydrolysis of flavonol glycosides apparently causes the predominant form in wine to be the aglycones, quercetin accounting for about 85% with

myricetin and kaempferol the remainder (1). Rutin, hesperidin,
and the flavanone glycoside naringin have been reported at
about 3 to 9 mg/kg each in the skin of a series of white grapes
(82).

The bitterness of naringin is nearly as great as quinine
with 6 mg/l easily detectable in water (8) and 300 mg/l or
more found in grapefruit juice. The bitter thresholds for
kaempferol, quercetin and myricetin in 5% aqueous ethanol
were 20, 10, and 10 mg/l and in beer over 50, 20, and 10 mg/l
respectively (50). Quercitrin, on the other hand, was reported
to be bitter, threshold 250 mg/l, in beer (42) and glycosides
are ordinarily more bitter than aglycones. The isoflavone
glycosides from subterranean clover were about 1/10 as bitter
at about 6-30 mg/l as isomolar quinine but the aglycones were
tasteless at the same level (83). The fully methoxylated
flavones of orange juice were bitter and had thresholds in a
synthetic medium like orange juice of 15-46 ppm (84).

The two major flavan-3-ols of wine are (+)-catechin and
(-)-epicatechin with small amounts of gallocatechin and gallate
derivatives (1, 63, 85). The total catechin group content in
white wines from juice is about 10-50 mg/l and up to perhaps
800 mg/l in red wines (1, 85, 86). Monomeric flavan-3,4-diols
do not appear to occur in grapes or wine, but polymeric forms
of anthocyanogens do. These condensed tannins can be converted
in poor yield to anthocyanidins, but they can be considered as
catechin or leucoanthocyanidin polymers since the usual linkage
involves the 4 position and catechin as well as cyanidin are
usual products of depolymerizing reactions.

The catechins usually account for much of any flavonoid
content in the lightest white wines with little or no leucoanthocyanidins present. As pomace contact is increased, both
the catechin and anthocyanogenic fractions increase, the polymeric tannins faster than the catechins. A white wine in the
high end of the usual total phenol range for white wine had
about equal anthocyanogen and catechin contents based on a differential browning versus reddening reaction, whereas in a red
wine the "anthocyanogen" was about 4-fold the apparent catechin
content (86). The method was developed for beer and its application to anthocyanin-bearing red wine is questionable without
further study, but this result seems reasonable.

If we estimate that typical red wine contains about 250
mg GAE/l as catechins then the remaining unassigned flavonoid
content should be a reasonable estimation of the dimeric and
larger condensed tannins. Thus a young red wine of 1400 mg
GAE/l of total phenol might be expected to have at least
200 mg GAE/l nonflavonoid, about 120 mg GAE/l anthocyanin
(calculated from 350 mg malvidin-3-glucoside), perhaps 50 mg
GAE/l flavonols, 250 mg GAE/l catechins and by difference
about 750 mg GAE/l as polymeric anthocyanogens. With a procedure involving precipitation with methyl cellulose of the
condensed tannins but not the monomeric flavonoids, the ap-

parent tannin content of a white wine having 337 mg GAE/1 total
phenol was 0-66 mg GAE/1 depending on the conditions. Red
wines with total phenols of 1560 and 1250 mg GAE/1 gave 273
and 488 mg GAE/1 of tannin (87). A different procedure also
designed to precipitate selectively the larger astringent
tannins precipitated 22 mg GAE/1 from a white wine of 350 mg
GAE/1 total phenol and 460 mg GAE/1 from a 1460 mg/1 total
phenol red wine (88). In general, about 1/3 of the total
phenol was precipitated from red wines. Similar data have
been obtained by other workers and with other techniques.

The fact that in commercial red wines, even of minimum
age, a majority of the anthocyanin pigment is incorporated
into the polymeric portion does not seem to invalidate the
general conclusions. Since the units with which the pigments
combine are already dimeric anthocyanogens or larger, they
would already have been included in the polymeric fraction
and the spectrophotometric determination of the total colored
anthocyanins should be reasonably correct even if the red
chromophores are incorporated into polymers. The red but
nondialyzable polymeric tannin from older red wine is still
astringent to taste, but whether anthocyanin incorporation
has affected flavor is not specifically known.

The molecular weight of the polymeric tannin in very
young wine is apparently of flavonoid dimer or trimer size
and increases to about 8-14 flavonoid units after a few years
of aging (MW 2-4000) eventually decreasing again in very old
wine (89).

The fact that astringency results from protein precipitation in the mouth and these tannins can be separated from wine
by precipitation with proteins (1, 21) indicates their importance in red wine flavor. The catechin and tannin flavors
we have studied together, specifically the fractions from
grape seeds which can be prepared in large amount without
contamination from anthocyanins.

(+)-Catechin has been reported to have a bitter taste in
5% ethanol or beer and thresholds in the two fluids of 20 mg/1
(50). The grape seed phenolic complex has been studied as such
and after fractionation by partition chromatography into
chromatographically distinct catechins (ether soluble), small
leucoanthocyanins (ethyl acetate soluble), larger anthocyanogenic tannins (butanol soluble), and the largest anthocyanogens
and phlobaphenes (retained on the column). These fractions
are designated respectively F, A, B, and O and contain the
major phenolic fractions indicated on the paper chromatogram
in figure 1 (63, 90).

These fractions had the approximate absolute thresholds
in water of 25 mg/1 for the whole phenol complex, 3.5 mg/1 for
the O+B group, and 20 mg/1 for the A or E (catechin) groups.
When added to a light white table wine all the fractions were
bitter and astringent at the levels tested except the catechin
fraction was not found to be astringent. The thresholds in

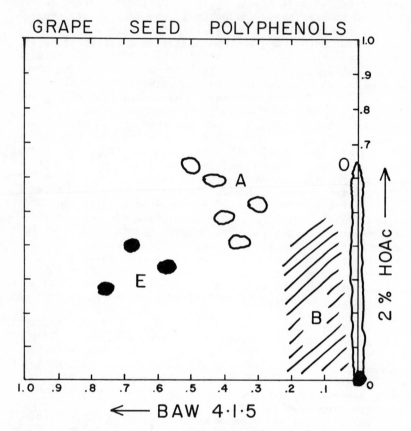

Figure 1.   Paper chromatogram of grape seed phenols. Fractions separated by column partition chromatography are: E, catechins (eluted by ether); A, smallest leucoanthocyanidins (eluted by ethyl acetate); B, leucoanthocyanidin oligomers (eluted by butanol); and O, tanninphlobaphenes (washed off with ethanol).

wine were for bitterness: E-200 mg/1, A-120 mg/1, O+B-12 mg/1, and the total complex 80 mg/1 (91). Note that the 80 mg/1 taste threshold for the seed complex is nearly identical with the 100 mg GAE/1 indicated earlier as the minimum increment for recognizable astringency increase from total pomace phenols. The same fractions had the same thresholds for astringency in white wine except the catechins lacked astringency and the O+B fraction astringency threshold, 20 mg/1, was nearly double its bitter threshold. It appears from these data that the phenols of white wine certainly in sum would produce bitterness and probably account for the bitterness level in these wines. In red wines the catechins alone would appear to be above the threshold for bitterness and the anthocyanogenic tannins would be of the order of 10 times the threshold level.

In on-going work (92), the average bitterness rating for grape seed phenol complex in a model wine by a trained panel was 4.7 at 55 mg GAE/1, 5.7 at 110 mg GAE/1 and 5.3 at 160 mg GAE/1 while the astringency increased regularly 2.4, 3.1, 3.4 with increasing phenol concentration. Both were evaluated on unstructured scales where 0 equalled no bitterness or astringency and 10 equalled extreme. These data are considered to show that sufficiently increased astringency masks bitterness as was also indicated in the pomace extraction data in table 1. This is an important observation because it explains the unmasking of bitterness which is sometimes observed as a wine ages and loses total tannin.

With the separate fractions, fraction A was significantly more astringent at 225 mg GAE/1 than at 175 mg GAE/1. The fraction O was significantly more astringent and more bitter per mg GAE than any of the other fractions. This may have special importance since this fraction seems to increase as an artifact of processing the grape seed complex. The mean bitterness ratings per unit mean astringency ratings were E-2.4, A-1.9, B-1.7, and O-2.2. Thus the catechin and the tannin-phlobaphene fractions appear more bitter and the leucoanthocyanidin fractions, particularly that soluble in butanol, more astringent, although all fractions exhibited both bitterness and astringency.

Pertinent comparisons can be made with the excellent work of Lea and Timberlake (93) on the phenols of cider. They found that oxidation decreased the astringency and/or bitterness and that most of the astringency was associated with units more complex than trimeric anthocyanogens.

VII. Indirect Flavor Effects of Phenols. The bitter and astringent phenols have interacting effects on other sensory characters of wine, but they have not yet been well studied.

Astringency has been shown to affect bitterness, and mutual
interactions can also be expected with sweetness, acidity,
alcohol level, etc. These interactions plus the direct and
additive effects of the various phenols in wine certainly
indicate the importance of phenols in wine flavor and the
need for more study.

The indirect chemical effects are also very large
between phenols and flavor. Only two examples will be
mentioned. It has recently been shown that the autoxidation
of phenols in wine solution leads to the coupled oxidation
of ethanol to acetaldehyde (94). This oxidation does not
occur without the phenol under the conditions pertaining in
wine. It occurs because a strong oxidant is cogenerated as
the phenol oxidizes. This system is believed to have considerable importance in flavor and other changes during wine
aging.

A specific type of bitterness is sometimes generated
in wine by the action of bacteria. This has been linked to
the formation of acrolein, but acrolein added to wine or
distillate does not have the same effect (82, 95). It
appears that the bitterness results from reaction of the
acrolein and perhaps other aldehydes with the phenols of
wine.

Abstract

Phenols other than flavonoids are similar in all wines
and make up most of the phenols in light white wines, about
200 mg/l calculated as gallic acid. Volatile phenols, perhaps 5 mg/l, appear to be below their individual flavor
thresholds in most wines, but are probably additive to give
sensory effects as a group. Tyrosol (15 mg/l), hydroxy-
benzoates (30 mg/l) and hydroxycinnamates (150 mg/l) are
individually near or below estimated threshold flavor levels,
but have similar bitter or harsh effects and probably are
additively important to wine flavor. Flavonoids have large
flavor effects in red wines. Typical young red table wines
have, as gallic acid, about 120 mg/l anthocyanins, 50 mg/l
flavonols, perhaps 5 mg/l flavanones, 250 mg/l catechins,
and 750 mg/l anthocyanogenic tannins. This tannin level is
5-10 times the flavor threshold, contributing bitterness and
astringency very important to wine quality. The other
flavonoid groups appear to be close to their flavor threshold
concentrations, and together certainly make direct and indirect contributions to wine quality.

Literature Cited

1. Singleton, V. L., Esau, P., "Phenolic Substances in Grapes and Wine, and their Significance," Advan. Food Res. Suppl. 1, Academic Press, New York, 1969.
2. Singleton, V. L., "The Chemistry of Plant Pigments," C. O. Chichester, Ed., p. 143, Advan. Food Res., Suppl. 3, Academic Press, New York, 1972.
3. Amerine, M. A., Berg, H. W., Cruess, W. V., "The Technology of Winemaking," 3d Ed., Avi Publ. Co., Westport, Conn., 1972.
4. Moncrieff, R. W., "The Chemical Senses," 3d Ed., CRC Press, Cleveland, Ohio, 1967.
5. Connell, D. W., Flavour Ind., (1970), 1, 677.
6. Joslyn, M. A., Goldstein, J., Advan. Food Res., (1964), 13, 197.
7. Shallenberger, R. S., Gustation Olfaction, Internatl. Symp., 1970, p. 126, O. Guenther, Ed., Academic Press, New York, 1971.
8. Horowitz, R. M., "Biochemistry of Phenolic Compounds," J. B. Harborne, Ed., p. 545, Academic Press, New York, 1964.
9. Horowitz, R. M., Gentili, B., Sweetness Sweetners, Proc. Ind.-Univ. Co-op. Symp. (1971), 69, G. G. Birch, Ed., Applied Sci. Publ., London.
10. Birch, G. G., Lee, C. K., Lindley, M. G., Staerke, (1975), 27, 51.
11. Palter, R., Lundin, R. E., Phytochemistry, (1970), 9, 2407.
12. Herrmann, K., Deut. Lebensm.-Rundschau, (1972), 68, 105, 139.
13. Singleton, V. L., "Chemistry of Winemaking," A. D. Webb, Ed., Advan. Chem., (1974), 137, 184(a), 254(b).
14. Singleton, V. L., Rossi, J. A., Jr., Amer. J. Enol. Viticult., (1965), 16, 144.
15. Singleton, V. L., Amer. J. Enol. Viticult., (1966), 17, 126.
16. Singleton, V. L., Amer. J. Enol. Viticult., (1972), 23, 106.
17. Singleton, V. L., Draper, D. E., Amer. J. Enol. Viticult., (1964), 15, 34.
18. Aubert, S., Poux, C., Ann. Technol. Agr., (1969), 18, 111.
19. Bourzeix, M., Mourges, J., Aubert, S., Vigne et Vin, (1970), 4, 447.
20. Berg, H. W., Akiyoshi, M., Food Res., (1957), 4, 373.
21. Singleton, V. L., Tech. Quart. Master Brew. Ass. Amer., (1974), 11, 135.
22. Negre, E., Bull. O.I.V. (Office Internatl. Vigne Vin), (1942), 15, (154), 20; (1943), 16, (155), 25.

23. Wucherpfennig, K., Hadi Khadem, S., Hensel, R., Goergen, W., Weinberg Keller, (1972), 19, 449.
24. Dittrich, H. H., Sponholz, W. R., Kast, W., Vitis, (1974), 13, 36.
25. Wucherpfennig, K., Bretthauer, G., Wein-Wiss., (1968), 23, 174.
26. Flanzy, M., Aubert, S., Ann. Technol. Agr., (1969), 18, 27.
27. Ough, C. S., Amer. J. Enol. Viticult., (1969), 20, 93.
28. Singleton, V. L., Sieberhagen, H. A., DeWet, P., Van Wyk, C. J., Amer. J. Enol. Viticult., (1975), 26, 62.
29. Kramling, T. E., Singleton, V. L., Amer. J. Enol. Viticult., (1969), 20, 86.
30. Berger, W. G., Herrmann, K., Z. Lebensm.-Unters.-Forsch., (1971), 147, 1.
31. Singleton, V. L., Sullivan, A. R., Kramer, C., Amer. J. Enol. Viticult., (1971), 22, 161.
32. Dzhakhua, M. Ya., Drboglav, E. S., Dzhaparidze, M. A., Vinodel. Vinograd. SSSR, (1975), (1), 50.
33. Dubois, P. J., Brule, G., C. R. Acad. Sci., Ser. D., (1970), 271, (17), 1597.
34. Webb, A. D., Muller, C. J., Advan. Appl. Microbiol., (1972), 15, 75.
35. Fiddler, W., Parker, W. E., Wasserman, A. E., Doerr, R. C., J. Agr. Food Chem., (1967), 15, 757.
36. Hruza, D. E., Van Praag, M., Heinsohn, H., Jr., J. Agr. Food Chem., (1974), 22, 123.
37. Steinke, R. D., Paulson, M. C., J. Agr. Food Chem., (1964), 12, 381.
38. Brule, G., Ann. Technol. Agr., (1973), 22, 45.
39. Whiting, G. C., Carr, J. G., Nature, (1959), 184, 1427.
40. Whiting, G. C., Coggins, R. A., Antonie van Leeuwenhoek, (1971), 37, 33.
41. Williams, A. A., J. Inst. Brew., London, (1974), 80, 455.
42. Meilgaard, M. C., Odour and Taste Substances, Internatl. Symp., Proc., Bad Pyrmont, Fed. Rep. Germany, Oct. 2-4, 1974.
43. Wasserman, A. E., J. Food Sci., (1966), 31, 1005.
44. Daun, H., Lebensm.-Wiss. Technol., (1972), 5, 102.
45. Liebich, H. M., Koenig, W., Bayer, E., J. Chromatogr. Sci., (1970), 8, 527.
46. Nishimura, K., Masuda, M., J. Food Sci., (1971), 36, 819.
47. Steiner, K., Schweiz. Brau-Rundschau, (1968), 79, 176.
48. Tanner, H., Mitt. Geb. Lebensmittel-unters. Hyg., (1972), 63, 60.
49. Stahl, W. H., "Compilation of Odor and Taste Threshold Values Data," Amer. Soc. for Testing and Materials, Philadelphia, PA, Data Series DS 48, 1973.
50. Dadic, M., Belleau, G., Amer. Soc. Brew. Chem., Proc., (1973), 107.

51. Guymon, J. F., Crowell, E. A., Qual. Plant. Mat. Veg., (1968), 20, 320.
52. Grohmann, H., Muehlberger, F. H., Deut. Lebensm.-Rundschau., (1954), 50, 183.
53. Yamamoto, A., Nippon Nogeikagaku Kaishi, (1961), 35, 824.
54. Moutounet, M., Ann. Technol. Agr., (1969), 18, 249.
55. Sapis, J. C., Ribéreau-Gayon, P., Ann. Technol. Agr., (1969), 18, 221.
56. Cordonnier, R., Rev. Francaise Oenol., (1973), 14, (50), 15.
57. Charalambous, G., Bruckner, K. J., Hardwick, W. A., Weatherby, T. J., Tech. Quart. Master Brew. Ass. Amer., (1972), 9, 131.
58. Rosculet, G., Brewers Digest, (1971), 46, (6), 68.
59. Szlavko, C. M., J. Inst. Brew., London, (1973), 79, 283.
60. Mitsuaki, M., J. Fermentation Technol., (1953), 31, 424, 495.
61. Horwitz, D., Lovenberg, W., Engelman, K., Sjoerdsma, A., J. Amer. Med. Assoc., (1964), 188, 1108.
62. Cerruti, G., Remondi, L., Riv. Viticolt. Enol. (Conegliano), (1972), 25, 66, 451.
63. Su, C. T., Singleton, V. L., Phytochemistry, (1969), 8, 1553.
64. Maga, J. A., Lorenz, K., Cereal Sci. Today, (1973), 18, 326.
65. Fantozzi, P., Bergeret, J., Ind. Aliment Agr., (1973), 90, 731.
66. Rapp, A., Ziegler, A., Vitis, (1973), 12, 226.
67. Noguchi, M., Yamashita, M., Arai, S., Fujimaki, M., J. Food Sci., (1975), 40, 367.
68. Mondy, N. I., Metcalf, C., Plaisted, R. L., J. Food Sci., (1971), 36 459.
69. Sanderson, G. W., Ger. Offen., (1974), 2,425,592; Chem. Abst., (1975), 82, 123615.
70. Yamada, K., Tanaka, T., Ger. Offen., (1972), 2,224,100; Chem. Abst., (1973), 78, 56443.
71. Yamamoto, A., Sasaki, K., Sanuno, R., Nippon Nogeikagaku Kaishi, (1961), 35, 715.
72. Hennig, K., Burkhardt, R., Weinberg Keller, (1972), 9, 223.
73. Marche, M., Joseph, E., Goizet, A., Audebert, J., Lafon, J., Rev. Francaise Oenol., (1975), 15 (57), 1.
74. Haeseler, G., Misselhorn, K., Underberg, P. G., Branntweinwirtschaft, (1972), 112, 206.
75. Joseph, E., Marche, M., Connaissance Vigne Vin, (1972), 6 (3), 273.
76. Joseph, E., Marche, M., Rev. Le Paysan, (1973), (659), 14; (660) 12; (661) 20.
77. Bourzeix, M., Rev. Francaise Oenol., (1973), 14 (49), 15.

78. Margheri, G., Falcieri, E., Vini Ital. (1972), 14, 501.
79. Jurd, L., "The Chemistry of Plant Pigments," C. O. Chichester, Ed., p. 123, Advan. Food Res. Suppl. 3, Academic Press, New York, 1972.
80. Chiriboga, C. D., Francis, F. J., J. Food Sci., (1973), 38, 464.
81. Somers, T. C., Evans, M. E., J. Sci. Food Agr., (1974), 25, 1369.
82. Drawert, F., Bull. O.I.V. (Office Internatl. Vigne Vin), (1970), 43 (467), 19.
83. Francis, C. M., J. Sci. Food Agr., (1973), 24, 1235.
84. Veldhuis, M. K., Swift, L. J., Scott, W. C., J. Agr. Food Chem., (1970), 18, 590.
85. Berger, W. G., Herrmann, K., Z. Lebensm.-Unters.-Forsch., (1971), 146, 275.
86. Dadic, M., J. Ass. Off. Anal. Chem., (1974), 57, 323.
87. Montedoro, G., Fantozzi, P., Lebensm.-Wiss. Technol., (1974), 7, 155.
88. Mitjavila, S., Schiavon, M., Derache, R., Ann. Technol. Agr., (1971), 20, 335.
89. Ribéreau-Gayon, P., "Chemistry of Winemaking," A. D. Webb, Ed., (1974), Advan. Chem. 137, 50.
90. Singleton, V. L., Draper, D. E., Rossi, J. A., Jr., Amer. J. Enol. Viticult., (1966), 17, 206.
91. Rossi, J. A., Jr., Singleton, V. L., Amer. J. Enol. Viticult., (1966), 17, 240.
92. Arnold, R. A., Noble, A. C., 26th Annual Meeting Amer. Soc. Enologists, San Francisco, June 26-29, 1975, Paper 81.
93. Lea, A. G. H., Timberlake, C. F., J. Sci. Food Agr., (1974), 25, 1537.
94. Wildenradt, H. E., Singleton, V. L., Amer. J. Enol. Viticult., (1974), 25, 119.
95. Dubois, P., Parfait, A., Dekimpe, J., Ann. Technol. Agr., (1973), 22, 131.

Surveillance and Control of Phenolic Tastes and Odors in Water to Prevent Their Effects on Taste and Flavor of Foods

B. F. WILLEY

Water Purification Laboratory, City of Chicago,
1000 East Ohio St., Chicago, Ill. 60611

Cool, sparkling clear, pure water is perhaps the most bland, essentially tasteless yet delightfully refreshing beverage available to man. Safe potable water in ample quantity is necessary to man's very existence. He needs it physiologically, from a sanitation (cleanliness) standpoint and also to make the products he uses and the foods and beverages he consumes. Since it is so bland, so nicely tasteless, it becomes extremely sensitive to any substances having taste or odor properties, whether pleasant or unpleasant. It is unfortunate, however, that our appreciation for a safe, good tasting water supply has not caused us to be vigilant in protecting the quality of the sources from whence comes our palatable and potable water. Domestic and industrial discharges have been carelessly allowed to enter our raw water sources, polluting them to greater or lesser extent with toxic metals and organic compounds. The resulting pollutionary changes has given water utility people an ever increasing problem in providing consumers with a satisfactory product at reasonable prices.

Water utilities which use ground water sources (well waters) have been more fortunate than those which must use surface waters. Although the towns having well water are greater in number than cities using surface water, the percentage of the population which must rely on surface supplies is much higher. Most urban areas, where the bulk of our population lives, are served by utilities which use and treat surface waters to supply their customer's needs. (Figure 1.)   It follows then, that most citizens use, drink and prepare their food with water subject to contamination and which must be carefully treated to remove these potential contaminants. Unwanted organics can be either man made or naturally occurring. Both types can cause tastes or odors or in some other manner affect the quality of a domestic supply.

Tastes and Odors in Water.
      According to psychologists there are only four true taste

sensations; sour, sweet,salty and bitter. Dissolved inorganic
salts of copper, iron, manganese, potassium, sodium and zinc can
be detected by taste. Concentrations producing taste are stated
to range from a few tenths to several hundred milligrams per
liter of water (1). All other sensations ascribed to the sense
of taste are actually odors, even though the sensation is not
noticed until the material is taken into the mouth. In contrast
to the quantities of material required to provide a true taste
sensation, those which can be classified as odorous materials
are detectable when present in only a few micrograms per liter.
Most of these are also of organic origin and include the phenols
and phenol-like substances.

Phenol in Water and Its Removal.

Of the organic pollutants which affect the palatability of
drinking water, phenol has been one of those most extensively
studied. Baylis (2), Burtshell(3), Snoeyink (4) and Lee (5)
have led the way to proper use of chlorine and activated carbon
for complete removal and/or destruction of the odorous chlorophenols. Carol J. Murin summarized the work of the above listed
authors and the chemistry involved in the development of chloro-
phenolic compounds in a paper recently released by the University
of Illinois, Department of Civil Engineering. In it he states,
"The chlorination of phenol according to Burtshell, et al., proceeds stepwise with substitution of the 2, 4 and 6 positions of
the aromatic ring. Chlorine can react to form 2⁻ or 4⁻ chlorophenol and secondarily 2, 4⁻ or 2, 6 dichlorophenol and ultimately 2,4,6⁻ trichlorophenol. Finally, and most importantly
excess free available chlorine then reacts with 2,4, 6 trichlorophenol (T.O.1000) to form a mixture of nonphenolic and essentially non-odorous oxidation products."
Lee has shown that the formation of chlorophenols and their
ultimate destruction involves complex kinetics with as much as
eight interdependent reactions occurring simultaneously. Ultimately phenol destruction occurs when chlorine dosage is 6-20
times the phenol concentration with 1.5 to 5 hours of contact
time and a pH range of 7-8. Chicago provides 8 hours contact
time from plant inlet to outlet, all of which is maintained with
free chlorine residual between 0.4 to 0.8 mg/l. The pH values
range between 7.8 to 8.3. Therefore, the chlorophenols have
been destroyed long before the finished water leaves the plants.
Other cities with lesser contact times are not so fortunate and
undoubtedly must rely more heavily on activated carbon to remove any residual chlorophenols.

Other Taste and Odor Sources.

Phenolics are not the only chemicals which produce undesir-
able tastes or odors in potable water supplies. One type which
has provided Chicago with some of its worst problems is classified as CH or Chemical Hydrocarbon odor arising from refinery

wastes or oil slicks from barges or ship ballast tanks. Although these have been reduced in both number and intensity, we
will experience CH odor episodes several times a year, especially during December and January. Like phenols, they are easily
but not economically removed with powdered carbon.

Perhaps the most persistent source of taste and odor
troubles are those due to certain plankton (6) and to the Actinomycetes, molds and fungi. I can say persistent because
they can occur all around the year and problems do not relate
to the same organisms all of the time. The organism counts will
vary depending on lake conditions and temperature, with different organisms reaching their peak counts at different times
during the year. Among the plankton which cause odors are the
Blue-green algae (Cyanophyta), most of the Pigmented Flagellates,
especially Dinobryon, (Figure 2) and even some of the Diatoms.
The nature of the odor sensation is related to whether the
growths are moderate or abundant (2). Molds and fungi have
typical odors which are handled rather easily in the treatment
process. The problem which can exist with these organisms, however, is the necessity to be sure the spores are killed along
with the parent growth. This requires sufficiently high chlorine dosages, together with adequate contact time. If the
spores are not killed they may find haven and nutrients in dead
end mains producing local pockets of odor bearing water (7).

Perhaps the most significant odor producing organisms encountered in Lake Michigan are the Actinomycetes. Four have
been isolated, speciated using Electron Microscopy, and have
been individually cultured providing us with opportunity to
identify specific odors produced by each pure culture (8).
(Figures 3 through 8). Streptomyces longispororuber, S. grie-
seus and S. lavendulae plus one more Streptomyces have been i-
dentified in Lake Michigan water, also the Actinomyces of the
genus Micromonospora (Figure 9.) are found sporadically, more
often from late spring to early fall.

Actinos present a special problem since they produce one or
more water-soluble, odoriferous chemical substances which can
persist even after the organism itself is killed and removed
from the water by sedimentation and filtration. The most well
known of these compounds is a chemical called Geosmin which has
a T.O. Number in excess of 10,000. We have not been able to
grow or identify Actinos from treated water samples which have,
on occasion, contained an identifiable musty-moldy odor, characteristic of that produced by growing Actino cultures. The
odor was very similar to Geosmin. Furthermore, the water leaving
the plant had no detectable odor except slightly chlorinous (1Cc
or 2Cc) but apparently developed a reaction product with
chlorine out in the distribution system. Once detected, it was
corrected by increased activated carbon treatment at the plant.

Analysis of Water for Taste and Odor Substances.

Figure 1. *Central water filtration plant, world's largest (1964)*

Figure 2. *Dinobryon, a pigmented flagellate producing a fishy odor*

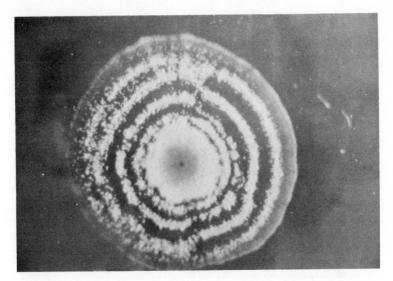

Figure 3.   *Light microscope photomicrograph of Streptomyces lavendulae. Note "fried egg" or halo effect.*

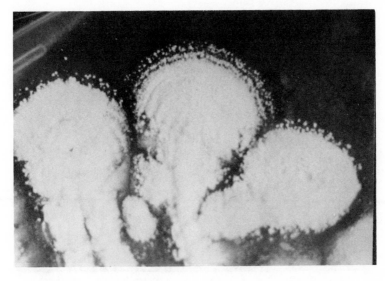

Figure 4.   *Light microscope photomicrograph of S. griseus "fan" pattern*

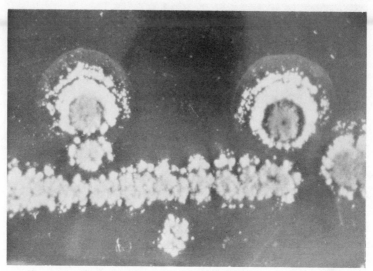

Figure 5.   Light microscope photomicrograph of S. longisporo-ruber

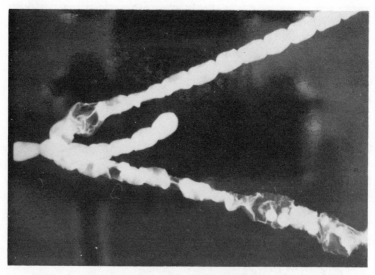

Figure 6.   Electron microscope photomicrograph of S. longisporo-ruber. Note straight chain and smooth spores.

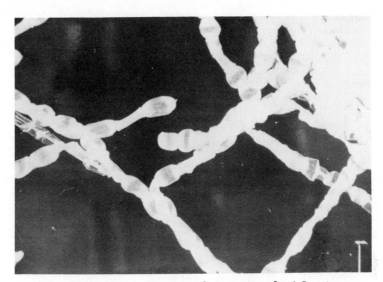

Figure 7. *Electron microscope photomicrograph of S. griseus.*
Spores smooth tear-drop shape spores with dark internal bands
(fimbria).

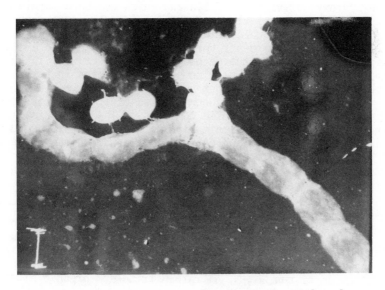

Figure 8. *Electron microscope photomicrograph of S. lavendu-*
lae. Note spores have spines and hairy appearance.

Although a few specific compounds responsible for taste and odor problems have become known, most are still unidentified. The analytical approach therefore is largely of an empirical nature. The human nose is still the best instrument we have, and final proof of effective treatment is based on either of two methods:

1. Threshold odor panels or teams, or,
2. Olfactory monitoring devices

1. Threshold Odors

The threshold odor number of a water is determined by diluting a sample of water with odor-free water until a dilution is found which is of the least definitely perceptible odor to the tester, or to each tester of a panel. For greatest accuracy a test panel of 5 to 10 persons is recommended. For day to day work in a water plant, however, two persons, who become quite experienced and capable, are usually used for routine odor testing. Odors are run both at ambient temperature and at 60°C (140°F). The latter permits detection of odors that might otherwise be missed. (Figure 10).

2. Odor Monitors

Odor Monitors are also used in the plant and provide for frequent comparisons of influent and finished water. Detection of any marked increase in raw water odor by the control lab chemists is reported immediately to the Filtration Engineer in charge and adjustments in treatment are made. (Figures 11 and 12).

Phenols and Phenol-Like Substances.

The term "phenols" as used in Standard Methods includes those hydroxy derivatives of benzene which can be determined using one or more variations of the 4-aminoantipyrene method prescribed in the manual. Limit of detection is approximately 1 ug/l. Phenolic compounds vary in their reactions to amino antipyrene, therefore, the method is not perfect. It is suitable, however, to determine a number of phenolics in terms of phenol which is used as the standard. (Figure 13).

Gas chromatography has been applied effectively for determination of a number of phenolic compounds. The method (still "tenative") uses a flame-ionization detector and three columns. Full directions with 12 suggested standards are given in section 222 F of the 13th Edition of Standard Methods. It has been adopted from ASTM D-2580-68 Standard, see ASTM Book of Standards, Part 23 (1968).

Nitrogen Compounds.

Nitrogen determinations such as NH_3-N or organic nitrogen can be helpful in indicating the presence of pollution of a water supply. Odorous substances may accompany increases in ammonia nitrogen or organic nitrogen readings. We have found it advantageous to run NH_3-N, Nitrate-Nitrite and organic nitrogen

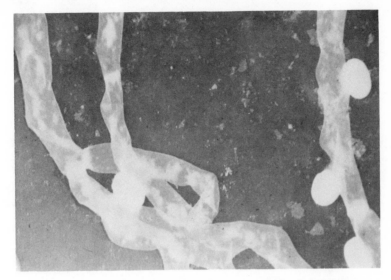

Figure 9. Electron microscope photomicrograph of Actinomycetes, genus Micromonospora. Does not produce pronounced odors and has lateral single spores.

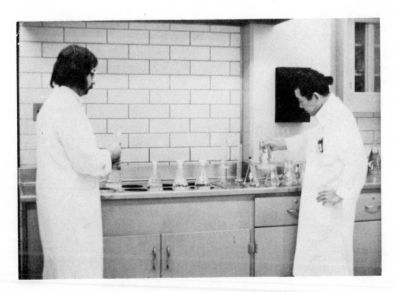

Figure 10.   Chemists making threshold odor tests

Figure 11.   Continuous-flow odor monitors, note spray

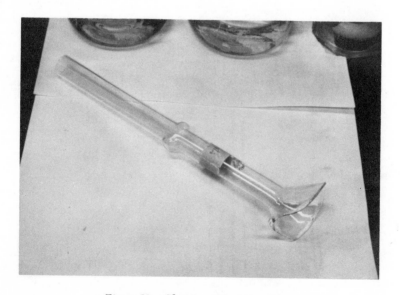

Figure 12.   Closeup view of osmoscope

as regular surveillance parameters on raw water at both crib and
shore intakes. Methodology used is covered in Standard Methods,
13th Edition; sections 132 through 135.

Treatment of Water for Taste and Odor Removal.
        The treatment process in Chicago consists of superchlorina-
tion, followed by alum coagulation, the addition of activated
carbon when required, lime to adjust pH, sedimentation, filtration, post chlorination and final pH adjustment with caustic
soda. Fluoride, in small quantity, is also added to reduce
dental caries.
        It was stated earlier that most odor problems result from
soluble organic matter present in the water supply. Since we
have learned much about chlorine chemistry and the types of
activated carbons available for water plant use, techniques have
been developed which permit removal of almost all odorous organics by proper sequential application of these two chemicals
during the coagulation, sedimentation and filtration process.
Superchlorination well beyond the breakpoint followed by powdered activated carbon during the coagulation process removes organisms, turbidity and adsorbable organics, mostly during sedimentation, with final polishing occurring as the water passes
through clean filters into the clearwells.
        Previously, at much lower chlorine dosages, chlorophenols
and other odorous reaction products were formed which were not
readily adsorbed by activated carbon. At that time it was
necessary to use ammonia or ammonium sulfate to tie up the
chlorine as chloramine to avoid chlorophenol odors. Superchlorination overcomes this problem, providing the excess chlorine required to convert phenols rapidly through the mono, di-
and trichlorophenol steps and on to oxidative breakdown products,
non-odorous or easily adsorbed on activated carbon, as described
by Burtshell,et. al. (3). If the chlorine demand is low, superchlorination alone will destroy tastes and odors to the extent
that carbon may not be necessary. The quality of Lake Michigan
water has improved noticeably and now taste and odor incidents
are much easier to treat and control.
        In closing I would like to quote a paragraph from a paper
on the benefits of free chlorine residuals presented by the
writer at the 1974 conference of the American Water Works Association, since it pretty well summarizes present conditions and
improvements which have occurred.
        "Remarkable improvement has occurred in recent years in raw-
water quality of both plants. By 1972 phenols were usually <1
ug/1 and ammonia nitrogen rarely exceeded 0.01-0.02 mg/1. Organic nitrogen has varied from a minimum of 0.03-0.12 mg/1 except for two days in May of 1972 when the level reached a high
of 0.22 mg/1. None of these values are high enough to make
significant effect on chlorine demand, which is about 0.2 mg/1
before a chlorine residual begins to appear. No true breakpoint

Figure 13. Phenol distillation rack

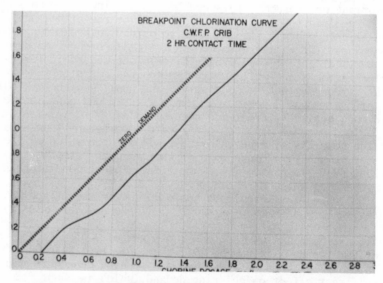

*Figure 14. Superchlorination curve for Lake Michigan water.
Note no breakpoint. Free residual occurs beyond 0.2 mg/l.
chlorine dosage.*

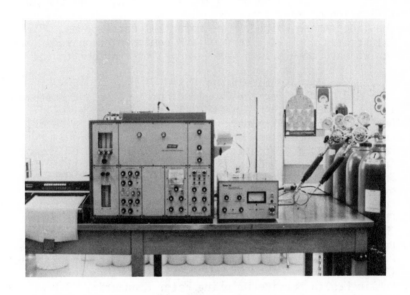

Figure 15. Tracor 550 gas chromatograph with Hall detector, used to determine low molecular weight, volatile, chlorinated hydrocarbons considered to be potential carcinogens.

occurs, as is shown by the curves in Figure 14.   Taste
and odor problems have virtually disappeared. Carbon feed has
been reduced from 365 days/year to a low at Central Plant of
only 50 days requiring carbon feed, during the entire year of
1973."

Work is still going on in the separation, isolation and i-
dentification of many trace organics, not only those which are
odorous, but those which may be potential carcinogens. Figure
15 illustrates   some of the equipment required. Although we have come a long way - we've lots farther to go!

Literature Cited

1. Standard Methods for the Examination of Water and Wastewater,
   13th Edition APHA, AWWA, WPCF, (1971).
2. Baylis, John R., Elimination of Taste and Odor in Water,
   Engineering Societies Monographs, (1935).
3. Burtshell, R. H., et. al., Chlorine Derivatives of Phenol
   Causing Taste and Odor., JAWWA, (51, 2, 205-14) (1959).
4. Snoeyink, V. L., et. al., Sorption of Phenol and Nitrophenol
   by Active Carbon, ES&T 3, 10, (918-24) (1969).
5. Lee, C. F., Kinetics of Reactions between Chlorine and
   Phenolic Compounds, Principles and Applications of Water
   Chemistry, Wiley and Sons, Inc. NY (1967).
6. Palmer, C. M., Algae in Water Supplies. US Dept. of H.E.& W.
   Public Health Service #657 - Taft Center, Cincinnati, Ohio
   (1962).
7. Public Health Service Drinking Water Standards, US Dept. of
   H.E. & W., Rockville, Md. (1962).
8. McMillan, L.M., et. al., Identification and Location of
   Actinomycetes in the Southern End of Lake Michigan, JAWWA
   64 (5) May, (1972).
9. Willey, R. E., et. al., Chicago's Switch to Free Chlorine
   Residuals, JAWWA, 67, (438), August (1975).

Simultaneous Detection of Nitrogen and Sulfur Containing Flavor Volatiles

DONALD A. WITHYCOMBE, JOHN P. WALRADT, and ANNE HRUZA

International Flavors and Fragrances, Inc., 1515 Highway 36, Union Beach, N. J. 07735

Element selective gas chromatography detectors are used extensively in the fields of pesticide and environmental chemistry. Their relative merits have been compared in recent reviews (1,2,3). A very recent paper by McLeod et al. (4) described the simultaneous operation of 5 detectors for pesticide residue analysis. The detectors which they used were: flame ionization (FID), flame photometric (FPD) at 526 nm for phosphorous, FPD at 394 nm for sulfur, Coulson electrolytic conductivity for nitrogen, and electron capture for halogen.

Our own system, which utilizes flame photometric (FPD) and alkali flame ionization (AFID) detectors for the analysis of sulfur and nitrogen containing compounds in addition to the FID, has been the subject of a short applications note (5) and will be described in detail here.

While there are several alternative detector types which are sensitive to nitrogen and/or sulfur containing compounds this discussion will be limited to those we consider to be the most useful. For sulfur selective response, the flame photometric detector (FPD) is by far the easiest and most reliable to use. Organics in the column effluent stream are burned in a hydrogen-rich flame in which they are fragmented and excited to higher energy states. The excited S_2 species formed from the combustion of sulfur containing compounds emits photons at a wavelength of 394 nm. The emitted light is passed through an appropriate interference filter and detected by a photomultiplier tube. The response to sulfur containing compounds is virtually free from interference by other compounds normally encountered in flavor research (selectivity is about 10,000:1). The sulfur response curve is peculiarly non-linear. Recently, electronic linearizing circuits

have been offered by instrument manufacturers for the
flame photometric sulfur detector.

The two detectors most often used for nitrogen are
the alkali flame ionization detector (AFID or therm-
ionic) and the electrolytic conductivity detector. The
AFID is the simplest in design, requiring only a source
of alkali metal as an addition to the usual flame ionization detector. The mechanism of enhanced response
to nitrogen and phosphorous compounds is still largely
unknown. In the most favorable cases, selectivity for
nitrogen containing components may approach 10,000:1.
However, in practice, selectivities on the order of 100
to 500:1 are more common.

The AFID is very sensitive to changes in the gas
mixture and position of the alkali metal salt source.
Altering the size or leanness of the flame changes the
temperature of the alkali metal source and can dramatically change the mode of response, selectivity, and
noise level. For example, the degree of response to
sulfur and halogen containing compounds when operating
as a nitrogen detector are controlled to a considerable
extent by the flame conditions. Recently, Perkin-Elmer
has introduced an improved AFID which they claim to be
virtually free of these effects.

The electrolytic conductivity detector (ECD) (6,
7), while a more complex device, is probably a more
reliable quantitative detector for nitrogen compounds
than the AFID. The GC effluent is pyrolyzed in a hydrogen atmosphere to produce NH_3 which is absorbed in
a water or ethanol stream and passed through an electrolytic conductivity cell for measurement. Halogen
response may be eliminated by using a basic absorber
after the pyrolysis unit. In the oxidative mode this
detector is sensitive to halogens and sulfur.

For the purposes of the largely qualitative survey type analyses which we perform, the improved ver-
sions of the AFID would be preferred over the ECD for
nitrogen detection from the standpoint of simplicity.
The flame photometric detector for sulfur compounds
has been extremely reliable in our hands and we see no
other detector comparable to it.

Instrument Design and Modifications

The primary objective in the design of our chromatographic system was the achievement of a totally
inert system in which even the most unstable compounds
could be delivered to the point of detection and/or
evaluation without decomposition. Many commercially
available instruments and systems reported in the

literature incorporate an essentially inert injection system and column, only to deliver the sample to the detector through a highly reactive hot metal transfer line. When chemicals are to be evaluated for use in flavors and fragrances, even the slightest decomposition of the chemical may lead to erroneous conclusions concerning its true aroma.

Although the chromatographic system which will be discussed is compatible with all types of columns, it has been specifically modified and optimized for use with large bore wall-coated glass capillary columns. We have found in our laboratory that many naturally occurring sulfur containing compounds are less subject to decomposition on a large bore glass capillary column than on a packed glass column.

The instrument which has been used in our laboratory is a Tracor MT-220 gas chromatograph equipped with a linear temperature programmer, three electrometers, dual flame ionization detectors (FID), and a Melpar flame photometric detector (FPD) (8) (Fig. 1). The Melpar flame photometric detector incorporates a flame ionization electrode which permits the simultaneous output of a flame photometric response and a flame ionization response. The FPD has been repositioned on the injector and detector manifold to accommodate the admission of the column effluent directly into the flame, as indicated by the glass-lined steel lines, thereby eliminating the possibility of contact between the sample gas and metal detector components prior to combustion.

One of the standard flame ionization detectors has been converted to an alkali flame ionization detector (AFID) (Fig. 2) by carefully positioning a small wedge-shaped segment of a 1.5 x 10 mm rubidium sulfate-potassium bromide pellet on the annular glass sleeve which surrounds the burner tip. The pellet is hydraulically pressed from a homogeneous 1:1 mixture of rubidium sulfate (99% Matheson Coleman & Bell) and infra red quality potassium bromide (Harshaw Chemical Co.) (9). The sample is combusted in a hydrogen-poor flame. When the flame has been properly optimized, the AFID provides an ionization current which is approximately 100 fold enhanced in response to nitrogen containing compounds.

A diagram illustrating the relationship of the various components of the system is shown in Figure 3. The injection port insert is constructed from a 15 cm segment of 6 mm o.d. x 1 mm i.d. capillary bore glass tubing. One end is slightly flared to facilitate direct syringe injection and the other end is drawn to

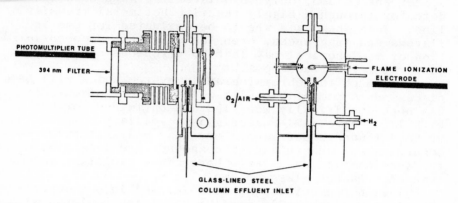

Figure 1. Schematic of Melpar flame photometric detector designed for simultaneous flame ionization and flame photometric (sulfur) operation with glass capillary columns

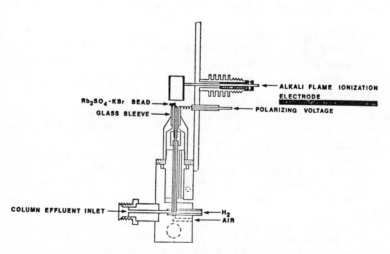

Figure 2. Schematic of a Tracor flame ionization detector modified for operation as an alkali flame ionization detector (nitrogen)

a tapered 1/16" o.d. tip. The 400' x 0.032" i.d. glass capillary column (C) is connected to the injection port insert and the exit splitter assémbly by small segments of heat-shrinkable Teflon tubing (B).

The column effluent is directed into one port of a 5-port effluent splitter assembly. The splitter is divided into two sections with all interconnecting lines fabricated from 1/16" o.d. x 0.5 mm i.d. glass lined steel tubing (Scientific Glass Engineering Pty. Ltd.) (A). The junction points are fabricated from low dead volume Swagelok unions, specially bored and fitted to minimize sample contact with stainless steel surfaces. To further reduce the possibility of sample decomposition, the intact splitter assembly was deactivated by coating with a 10% SE-30 solution and heating at 400°C according to the procedure of Thompson and Goode (10). The first section of the splitter assembly merely provides for the addition of helium make-up gas from one of the auxiliary injection port flow lines (D). The diluted effluent is then directed to the 3-way splitter of the second section. At this point, 20% of the effluent is directed to the FID/FPD, 30% to the AFID, and 50% to a heated exit port. The signals from the three detectors are fed into three electrometer channels which are connected to a Soltec/Rikadenki three-channel recorder. The FID signal is connected in parallel to a central GC computer system which provides a quantitative estimation of the sample components and calculates their retention index (I_E) (11).

The heated exit port is a modification of a third injection port which provides two options: 1) sample collection for subsequent analytical routines, or 2) simultaneous odor evaluation of the eluting components. When used in the latter manner, additional helium is provided (E) concentric to the flow line to "lift" the components away from the hot surface. The chromatogram displaying the responses from the three detectors is simultaneously annotated to provide an "aroma profile" of the chromatographically resolved sample.

Detector selectivity is demonstrated in Figure 4. Shown is the separation of a model system composed of a number of sulfur and/or nitrogen containing compounds. The top trace is the FID response. As indicated in the second trace, the FPD responds exclusively to components 2, methyl ethyl disulfide; 5, 2-n-pentyl thiophene; and 8, tetrahydrothiophene-3-one. The AFID responds to components 3, 1-n-pentyl pyrrole; 6, 2-methyl-5-vinyl pyridine, and 10, n-pentyl pyrazine. Dual responses are observed on the FPD and AFID for

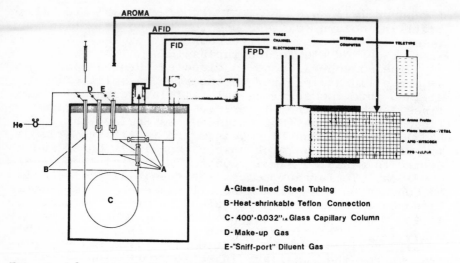

*Figure 3.   Schematic of the Tracor MT-220 gas chromatograph equipped with a 400' ×
0.032" id glass capillary column and modified for simultaneous detection and aroma evaluation of nitrogen and sulfur containing compounds*

A-Glass-lined Steel Tubing

B-Heat-shrinkable Teflon Connection

C- 400'·0.032"·ı Glass Capillary Column

D-Make-up Gas

E-"Sniff-port" Diluent Gas

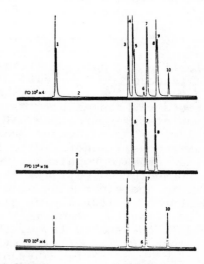

*Figure 4.   Chromatogram of synthetic
mixture using 3-detector chromatographic system*

1	ETHANOL	6	2-METHYL-5-VINYL PYRIDINE
2	METHYLETHYLDISULFIDE	7	4-METHYL-5-VINYL THIAZOLE
3	1-N-PENTYLPYRROLE	8	TETRAHYDROTHIOPHENE-3-ONE
4	FURFURAL	9	5-METHYL FURFURAL
5	2-N-PENTYLTHIOPHENE	10	N-PENTYLPYRAZINE

component 7, 4-methyl-5-vinyl thiazole, as would be expected. Also observed is a relatively small response of the AFID to the large quantity of solvent, ethanol. This sympathetic response can readily be distinguished from a true response by comparing the FID and AFID recorder traces and coincidence times. Under normal operating conditions the detector sensitivity is adjusted to give comparably sized peaks for all three detectors. The useable sensitivity (considering the S/N ratio) of this AFID is of the same order of magnitude as the FID; however, the FPD may achieve 10 to 100 times greater sensitivity.

The principle features and advantages of this multiple selective detector system may be summarized in the following three points:

1. An Inert, High Capacity, High Resolution Chromatographic System. The column systems developed permit the resolution of 1 to 2 microliter direct injections of solvent-free natural extracts while maintaining column efficiences on the order of 0.5 to 1.0 x 10^6 theoretical plates. The use of wide bore capillary columns is essential since the sample capacity of narrow bore columns with injection splitters is not sufficient for use with multiple effluent splitting. Our columns provide sufficient capacity to allow preparative isolation of microgram quantities of material for further spectral characterization.

2. Simultaneous Nitrogen, Sulfur, Flame Ionization, and Physiological Detection. The sensory evaluation of GLC effluents has been recognized as an aid in assessing the significance of the various components to the aroma of the total sample. When sensory evaluations are made in parallel with a FID, aromas and aroma changes are detected in areas which cannot be assigned specifically to a change in detector response. However, when multiple selective detectors are used, these aromas may often be assigned specifically to trace or unresolved nitrogen and/or sulfur containing components, many of which exhibit extremely low odor thresholds. This capability provides the flavorist, perfumer, or analyst with the ability to better characterize the sample and locate components of specific interest.

3. Provide Maximum Information for the Acquisition and Interpretation of GC-MS Data. In the same manner as above, it is possible to effectively locate trace and unresolved components prior to analysis by

gas chromatography - mass spectrometry (GC-MS) and, through the aid of computerized MS data handling facilities, obtain identifiable mass spectra of components which would normally be overlooked. When this technique is used in conjunction with specific or multiple ion monitoring of the mass spectral data the analyst is armed with powerful ancillary techniques for the location and mass spectral characterization of organic flavor and aroma components.

Application

In our laboratory natural extracts are routinely surveyed with the system just described before GC-MS analysis. An analysis of the volatile components of pressure cooked pork liver, which has previously been reported in part by Mussinan and Walradt (12) will provide an example of the application of the technique. The details of the sample preparation are essentially the same as previously reported, the product being a concentrated ether solution of cooked pork liver volatiles. The three chromatograms shown in Figure 5 are those of a 2 μl injection of the extract on a 400' x 0.032" i.d. glass SE-30 capillary column, temperature programmed from 50-190°C at 1°C/minute. The upper trace is the flame ionization response. The second trace is the AFID response to nitrogen components and the flame photometric response to sulfur compounds is the third trace. Due to the inability to suitably annotate the figure with the odor descriptors, they have been omitted.

The major component of the sample, 1, is furfuryl alcohol. The response of the AFID to the pyrazines and pyrrole derivatives compared to other constituents of similar FID response illustrates the detector selectivity as in the case of components 2 (methyl pyrazine), 3 (2,5(6)-dimethyl pyrazine), 4 (2-ethyl-6-methyl pyrazine) and 5 (trimethyl pyrazine). The information obtained from the flame photometric detector likewise enhances the confidence in the location and identification of sulfur containing components. The response of the FPD to components 6 (methyl mercaptan), 7 (dimethyl disulfide), and 8 (dimethyl trisulfide) is obvious. Benzothiazole, 9, is dramatically highlighted by the dual response of the AFID and FPD. The identification of component 10, 2-acetyl thiazole, may be readily attributed to the selective detectors in that the FID response is only slightly greater than the background. Through the use of the selective detectors and the retention index, increased confidence was gained in the

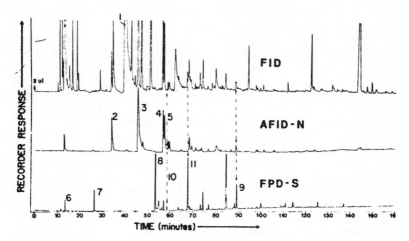

Figure 5. Chromatogram of cooked pork liver volatiles showing selective detector responses to nitrogen (AFID-N) and sulfur (FPD-S) containing compounds. Conditions: 400' × 0.032'' id glass capillary SE-30 column programmed at 1°C/min from 50° to 190°C.

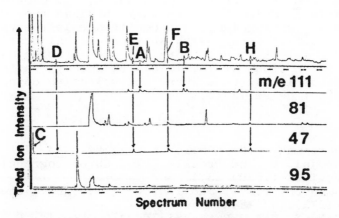

Figure 6. Computer regenerated total ion chromatogram of the GC–MS analysis of cooked pork liver volatiles with specific ion plots for masses (m/e) 111, 81, 47, and 45

mass spectral identification of this compound. The
methyl thiophene carboxaldehyde (11) is readily highlighted by the sulfur response.

More specific structural information may be obtained from a computerized GC-MS analysis. We are
presently using a Varian MAT-III GC-MS fitted with a
400' x 0.032" i.d. glass capillary column and inter-
faced to a Varian SpectroSystem 100-MS. Figure 6 de-
picts the total ion plot of a portion of the pork
liver analysis and, in addition, specific ion plots of
masses 111, 81, 47, and 95. This type of analysis is
an aid in differentiating specific classes of components from the more generalized element selective
analysis. In this case the mass 47 plot distinguishes
the alkyl mercaptans and sulfides, as exemplified by
methyl mercaptan (C), dimethyl disulfide (D), and dimethyl trisulfide (E). The location of acyl thiophenes may be highlighted by the use of the mass 111
plot as indicated by the thiophene carboxaldehyde (A),
and methyl thiophene carboxaldehyde (B). The system
may also be manipulated to provide specific indications of other classes of compounds not selected for
by other techniques, i.e., masses 81 and 95 plots for
alkyl and acyl furans, respectively.

While computerized GC-MS analysis can provide
more detailed structural information, the element selective detectors are a more economical and valuable
aid for rapidly locating and evaluating nitrogen and
sulfur compounds of interest in complex mixtures and
monitoring their presence through the necessary isolation and fractionation procedures. It provides information complementary to that obtained from GC-MS
analysis, giving increased confidence in the validity
of identifications. With appropriate calibration, the
element selective detectors can also provide quantitative estimations of low levels of specific heteroatomic
compounds in mixtures where non-selective detectors
would not be reliable.

Literature Cited

1. Hartmann, C.H., Anal. Chem. (1971) 43 113A.
2. Krejci, M. and Dressler, M., Chromatographic
    Reviews (1970) 13 1.
3. Natusch, D.F.S. and Thorpe, T.M., Anal. Chem.
    (1973) 45 1184A.
4. McLeod, H.A., Butterfield, A.G., Lewis, D.,
    Phillips, W.E.J. and Coffin, D.E., Anal. Chem.
    (1975) 47 674.

5. Walradt, J.P., in "Techniques of Combined Gas Chromatography/Mass Spectrometry: Applications in Organic Analysis", pp. 414-416, McFadden, W.H., Ed. Wiley. New York 1973.

6. Coulson, D.M., J. Gas Chrom. (1965) 3 134.

7. Hall, R.C., J. Chrom. Sci. (1974) 12 152.

8. Brody, S.S. and Chaney, J.E., J. Gas Chrom. (1966) 4 42.

9. Craven, D.A., Anal. Chem. (1970) 42 1679.

10. Thompson, C.J. and Goode, C.N., Science Tools (LKB Instrument Journal) (1972) 19 33.

11. van den Dool, H. and Kratz, P.D., J. Chromatog. (1963) 11 463.

12. Mussinan, C.J. and Walradt, J.P., J. Agr. Food Chem. (1974) 22 827.

6

Flavor Precursors in Food Stuffs

LEONARD SCHUTTE

UniMills B. V., Lindtsedijk 8, Zwijndrecht, The Netherlands

The Role of Food. Flavor is one of the properties
of a food by which it is recognized and enjoyed. The
properties may be subdivided into primary and secondary
ones. The primary function of food is nutrition or
fuel for the maintenance of life. Food should therefore be nourishing, have a proper nutritive value and
be devoid of toxic or harmful effects. Few materials
in nature fulfill these requirements entirely: some are
poisonous, others may contain enzymes that interfere
with the function of the body (biochemical poisons),
and still others are simply not digestible.

Animals are generally thought to know by instinct
what is good for them and what is not. In human beings
this instinct has been suppressed and replaced by another process - that of learning from experience. The
natural materials that, through the ages, have been
found to be nutritive and beneficial are fed to chil-
dren, who in turn become accustomed to them and continue to flourish on them. The same mechanism may also
apply to animals; what we call "instinct" may be no
more than a process whereby customs are passed on from
parents to offspring.

Years of experience have determined not only the
range of nutritious foods available, but also the method of treating them. For instance, natural materials
sometimes have to be heated to kill enzymes which have
adverse effects.

The nutritive effect and harmlessness of food have
been established by experience, however, this choice is
no longer determined on this basis. One does not eat a
natural material at random and wait to see whether it
does one good or makes one ill. Instead, the food is
recognized by its secondary properties. The process by
which memory and tradition link the secondary properties of the food is represented in Figure 1.

Beneficial or adverse physiological effects of
food are established by experience and can be correlated only indirectly with the primary properties.   On
the other hand, recognition of a food by its secondary
or sensory properties, which may be divided into appearance, texture and flavor, is direct.   Today, in
the case of packaged foods, "appearance" may be extended to the package and to the image created by ad-
vertisements, in which consumer's attention may be
called to the primary properties.   Such factors tend
to complicate the picture.

   Flavor in Food.   The preference for particular
combinations of secondary properties is known as per-
sonal liking, or "taste" in the broad sense of the
word.   The pleasure of eating is derived from these
secondary properties.   Taste is acquired, often over
generations.   This explains why eating is a social-
cultural phenomenon, not only the method of selecting
and preparing the food, but even the way of consuming
it, the so-called table manners.   This also explains
why the introduction of novel food products is so
difficult.   These products will only stand a chance of
success if they bear some resemblance to familiar foods
which appeal to the taste of the consumer.
   The secondary or sensory properties by which a
food product is judged are appearance, color, taste,
texture, temperature and in the case of popcorn and
chips for instance, even sound.   As shown in Figure 2,
these factors are closely connected and it is often
impossible to separate one from another (1).   They
determine whether, and to what extent, a particular
dish is acceptable to the consumer.   The secondary
properties are very important as people do not eat
nutrients, they eat food; the nutritional impact of a
balanced mixture of nutrients that is not accepted as
a food is nil.
   Taste in the broad sense is not very flexible,
but this does not mean that some tastes have not gradually changed.   For instance, in the recent past people in the United States and in Europe have grown accustomed to canned tomato soup and tomato juice, and
many prefer the "sterilized flavor" of these products
to fresh tomato flavor.   Fresh apple sauce is no
longer preferred to canned apple sauce.   Many of the
younger generation brought up on margarine do not like
the taste of butter.

Flavor in Food Systems

Taste Compounds. Although flavor is mainly associated with taste and odor, the other entities in
Figure 2 play an inseparable role. For instance, it
is very hard to identify banana flavor in a pink ice
cream. Flavor chemists, however, try to isolate those
factors imparting taste and odor from the other
entities. Consequently, flavor chemistry deals with
those compounds responsible for the taste and odor of
food. These flavor compounds can be subdivided into
two classes: taste compounds and odor compounds.
Their characteristics are summarized in Table I.

TABLE I

CHARACTERISTICS OF TASTE AND ODOR COMPOUNDS

Taste Compounds	Odor Compounds
Perceived by tongue	Perceived by nose
Non-Volatile	More or less volatile
Polar, water-soluble	More or less non-polar
Present in relatively high concentrations	Present in low concentrations (mg/kg or μg/kg range)
Salt, sweet, sour, bitter	Many different flavor sensations

One can perceive taste compounds on their own by
pinching the nose during eating or when one suffers
from a heavy cold. In this way the odor compounds are
prevented from reaching the nasal cavity, either directly or indirectly. Taste compounds are in general
non-volatile, water soluble and usually present in the
food in relatively high concentrations. They give a
combination of salt, sweet, sour and bitter im-
pressions. Examples are common salt, sucrose, citric
acid, caffeine and monosodium glutamate.

Odor Compounds. Odor compounds have a finite
vapor pressure. They are never completely polar or
non-polar, although some are more soluble in water and
others in organic solvents. They are usually present
in low concentrations, in the mg per kg or μg per kg
range, and they give a large variety of flavor sensations. In contrast to the non-volatiles, few volatiles are essential intermediates in the biochemistry
of the living organism. They may be considered as
being derived from the basic bulk ingredients of the

living organism, i.e. carbohydrates, proteins and fats
(triglycerides). The latter can thus be considered as
the precursors of the volatile flavor compounds. As
these bulk ingredients are similar in most living
species, it is not surprising that some odor compounds
are common to quite different groups of foods. Examples are furanones in fruits and meat, sulfides in
meat and vegetables, and pyrazines in vegetables and
baked or roasted products like bread and coffee.

The precursors, however, are present in different
proportions in the various living species and furthermore the mode of formation of the volatiles may vary
from one food to another. Possible modes of formation
are bioformation in the living species, enzymatic formation by flavor enzymes or in fermentation processes,
or non-enzymic Maillard-type reactions. The level at
which the various odor compounds are present, their
proportions and their mode of release determine the
overall flavor of the final food product.

In theory, all volatiles present in a food
product contribute to the overall odor, which makes
the flavor of every food product unique. Some compounds are, however, more important than others and
these are often referred to as "key compounds". It
would though be an over-simplification to think that
the flavor of a food product can be reconstituted by
proper combination of these key compounds. This has
been possible only in some exceptional cases, like
vanilla or cucumber.

The importance of flavor in determining the ac-
ceptability of food certainly warrants the study of
natural flavor compounds. Structural elucidation of
these compounds not only adds to our knowledge of the
types of molecules which produce flavor, but it also
supplies us with suitable materials for making highly
nutritious but less tasty foods more palatable. Moreover, flavoring can help to make acceptably priced,
attractive products more widely available. Knowledge
of the mechanism of flavor formation from precursors
is therefore not only of scientific interest. This
knowledge may be applied in selecting good flavoring
methods and in choosing processing conditions, so that
off-flavor formation is prevented. It is therefore
the task of the flavor chemist not only to isolate and
identify odoriferous compounds in food products, but
also to study their formation from precursors in order
to be able to apply the results to the flavoring of
processed and novel food products. This will be illustrated using meat flavor as an example.

Meat Flavor Compounds

Taste Compounds. There are, of course, many
types of meat flavors, e.g. boiled, brothy, roast,
etc. and beef, pork, lamb, chicken, etc. All these
flavors have many characteristics in common and I will
only distinguish between the various types when appropriate. The taste compounds in meat are rather
universal. The most important taste compounds in meat
are given in Table II. The natural pH of meat is
about 5.5, which should be borne in mind when assessing their contribution to the taste of meat.

TABLE II

SOME IMPORTANT TASTE COMPOUNDS IN MEAT FLAVOR
(pH VALUE 5.5)

glutamic acid
inosinic acid
succinic acid
lactic acid
phosphoric acid
pyrrolidonecarboxylic acid

Odor Compounds and their Precursors. The volatile part of meat flavor is more complicated. For the
sake of argument let us assume that an ideal gas chromatogram of meat flavor volatiles is as shown in
Figure 3. In this gas chromatogram all volatiles are
represented as separate peaks and the heights are pro-
portional to the relative amounts present in the headspace of the meat product concerned. By dividing the
concentration of each compound by its flavor threshold
value, an "aromagram" which shows the relative flavor
impression of each compound is obtained. This is
shown in Figure 4 which represents the hypothetical
aromagram of meat flavor, more or less comparable with
the skyline of a city.
    The flavor compounds in the aromagram of a cooked
meat product are formed from precursors present in the
raw meat. Precursors are those ingredients in the
food that are the most reactive during its preparation. Some sort of heat treatment is usually responsible for flavor formation in meat. The precursors of
meat flavor are certain amino acids and sugars which
react together in so-called Maillard reactions. Some
of these reactions will be briefly discussed. The
first reaction is the so-called Amadori rearrangement
(Figure 5) in which monosaccharides like glucose,

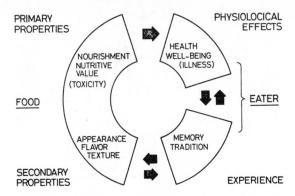

Figure 1. *Classical relations between primary and secondary properties of food and its effect on the eater*

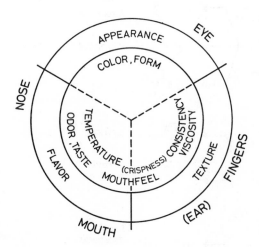

Figure 2. *Interconnection between sensory properties of food*

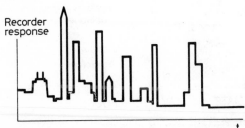

Figure 3. *Hypothetical gas chromatogram of meat flavor*

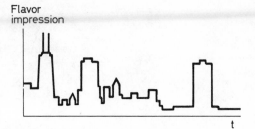

Figure 4.   *Hypothetical aromagram of meat flavor*

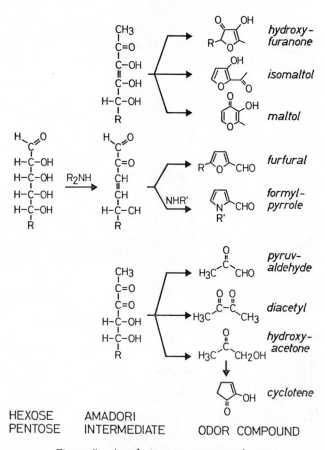

| HEXOSE | AMADORI | |
| PENTOSE | INTERMEDIATE | ODOR COMPOUND |

Figure 5.   *Amadori rearrangements of sugars*

fructose or ribose react with amines by which carbonyl groups are moved within the sugar molecule (2).   In subsequent cyclisation reactions heterocyclic compounds like furanones, pyranones or even pyrroles are formed. The rearranged sugar may also break down into smaller molecules like diketones or hydroxy acetone, which in turn may form a cyclotene in a condensation reaction. Another type of reaction is the Strecker degradation (Figure 6), in which aldehydes are formed from amino acids which have been oxidized by diketones. Examples are the formation of acetaldehyde from alanine, methional from methionine, phenylacetaldehyde from phenylalanine and benzaldehyde from phenylglycine. Strecker degradation of cysteine yields hydrogen sulfide, which itself is a meat flavor compound as well as being an important precursor of other sulfur-containing meat flavor compounds (3). For instance, reaction of hydrogen sulfide with heterocyclic compounds like furanones, leads to the formation of sulfur containing heterocycles which have a roasted meat flavor (Figure 7). Methanethiol, which is developed from methionine, is a similar compound, yielding mono-, di- and trisulfides. Some other compounds formed from these small precursor molecules are 1-methylthio-ethanethiol and dimethyltrithiolane.

Pyrazines are also formed as a result of Strecker degradations. Here the diketone moiety is held responsible for the direct formation of the pyrazines (4).

As shown in Figure 8 some reactions involve the formation of acylthiazoles from cysteine and diketones, such as pyruvaldehyde (3), and the formation of unsaturated lactones from α-ketobutyric acid, an oxidation product of threonine (5).

Figure 9 shows the flavor compounds formed from thiamine or vitamin B1. As the biosynthesis of the thiazole moiety of thiamine involves reaction of methionine with alanine, there is some connection between the formation of these flavor compounds and that of those formed directly from amino acids. It again indicates that the formation of meat flavor compounds may follow complicated pathways involving various intermediates.

The reactions discussed lead to well-identified flavor volatiles, which correspond to the prominent tall buildings in the skyline shown in Figure 4. They may be regarded as the "key compounds" or top-notes. In meat flavor such compounds are: carbonyl compounds, heterocycles, sulfur and nitrogen compounds and finally, some phenols. Some of them contribute more to

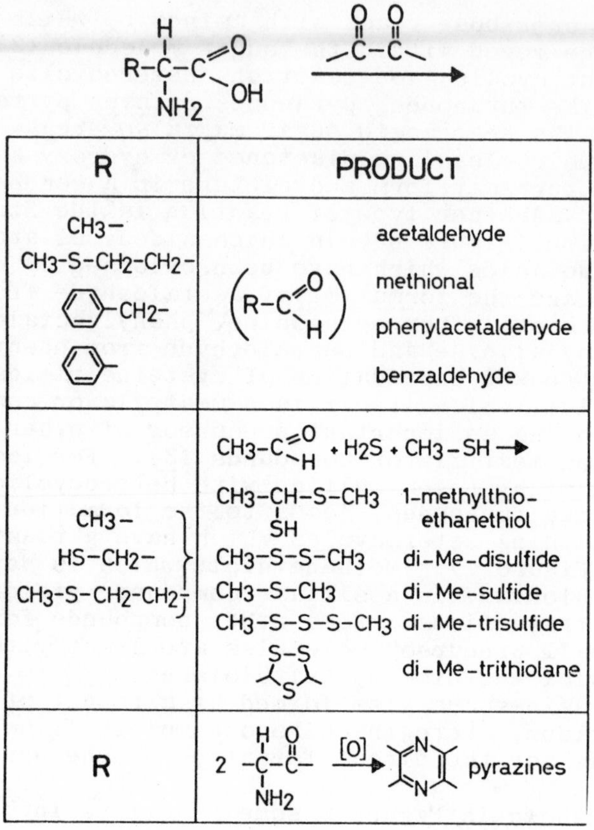

R	PRODUCT
CH_3- $CH_3-S-CH_2-CH_2-$ ⬡$-CH_2-$ ⬡$-$	$\left(R-C\overset{O}{\underset{H}{\diagdown}} \right)$ acetaldehyde methional phenylacetaldehyde benzaldehyde
CH_3- $HS-CH_2-$ $CH_3-S-CH_2-CH_2-$	$CH_3-C\overset{O}{\underset{H}{\diagup}}$ + H_2S + $CH_3-SH \longrightarrow$ $CH_3-CH-S-CH_3$ 1-methylthio- $\quad\quad\;SH$ ethanethiol $CH_3-S-S-CH_3$ di-Me-disulfide CH_3-S-CH_3 di-Me-sulfide $CH_3-S-S-S-CH_3$ di-Me-trisulfide (trithiolane ring) di-Me-trithiolane
R	$2\;\overset{H\;O}{\underset{NH_2}{-C-C-}}\;\xrightarrow{[O]}$ (pyrazine ring) pyrazines

Figure 6.   *Strecker degradation of amino acids*

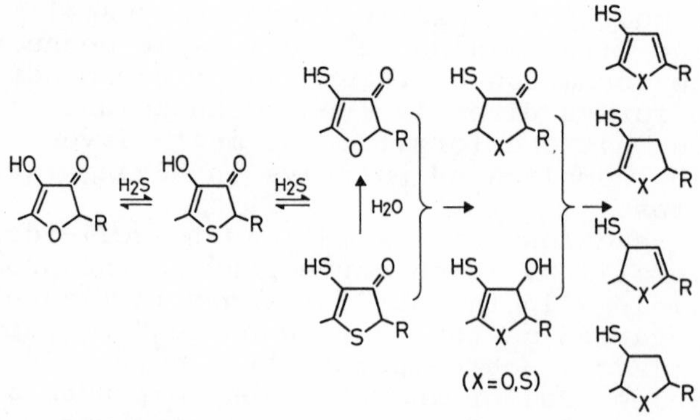

Figure 7.   *Flavor compounds from the reaction of hydrogen
sulfide with heterocyclic furanones*

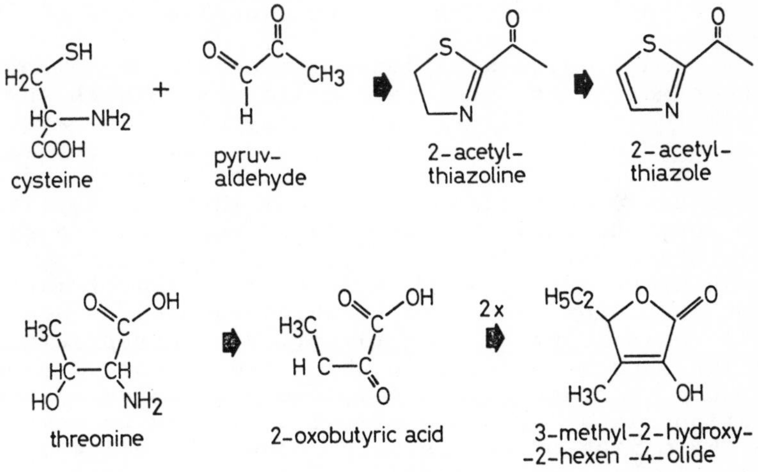

Figure 8.   *Flavor compounds derived from cysteine and threonine*

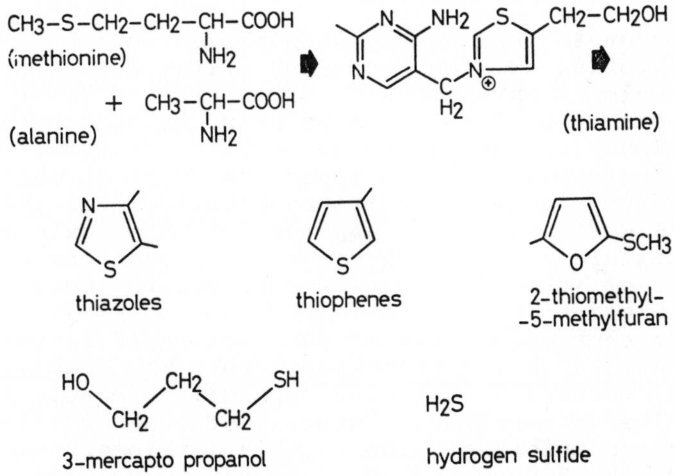

Figure 9.   *Flavor compounds from thiamine*

meat flavors than others and some may give a typical
character to the meat flavor, for instance, thiophenes
give a roasted character and decadienal a more chicken-
like flavor.

Although quite a variety of meat flavor compounds
has been mentioned, there are still many others that
contribute to a smaller degree, many of which have not
yet been identified. These compounds may be envisaged
as the small buildings in the skyline, giving a more
or less basic meat flavor. Without them the skyline
would seem unnatural and the flavor would have a synthetic character.

In other words, in a balanced meat flavor many
flavor compounds play a role, which themselves are not
at all reminiscent of meat flavor. An appropriate
name for such an effect, in which two or more flavor
compounds together give a completely different flavor
sensation, is a "synosmic effect". It may be compared
with the mixing of colors, e.g. blue and yellow together give a completely different color: green; or
with the sound of an orchestra, which is different
from that of any of the individual instruments. A
"synosmic effect" should be distinguished from an
"additive effect", flavors with the same flavor
character adding their intensities,and from a "synergistic effect", well-known in taste mixtures, where
one compound enhances the flavor sensation of another,
to a larger degree than in an additive effect.

Application of Meat Flavors

General. How can this knowledge be applied in
the flavoring of food products? First we must realize
what tools we have and secondly where and how these
tools can be used (6). If we start by looking at the
meat flavorings developed so far, we can first distinguish taste mixtures and taste enhancers in which
monosodium glutamate and ribonucleotides are the components most frequently applied. Then we have top-
note mixtures consisting of volatile odor compounds.
Release of these compounds may be controlled by micro-
encapsulation (7).

In some cases, use of derivatives of flavor volatiles with a labile "blocking" group may enable the
gradual formation, and hence gradual release, of the
volatiles in question. Examples of such artificial
flavor precursors are the O-t-alkyl thiocarbonates of
thiols, which upon heating form the corresponding
thiol flavor compounds (3). This is illustrated in
Figure 10.

Finally, there are the basic flavor mixtures which balance the overall meat odor. They are usually formed by letting suitable precursors react together. Protein hydrolysates are frequently employed for this purpose. The flavor of protein hydrolysates is formed by reactions between the amino acids and carbohydrates liberated during the hydrolysis of proteinaceous materials such as soy or gluten. More sophisticated precursor mixtures yielding meat flavors on heating have been widely patented (3) and (8).

As shown in Figure 11, the precursor systems generally consist of sulfur-containing compounds such as cysteine, thiamine or hydrogen sulfide on one hand, and carbonyl compounds like monosaccharides, aldehydes or furanones on the other. When such precursor systems are heated in water, boiled meat flavors are formed, whereas heating in fat gives rise to a meat flavor with a roasted character. The fat is also held responsible for the development of the typical flavor character of the various types of meat, such as chicken, pork or lamb. For instance, poly-unsaturated aldehydes and gamma lactones in chicken flavor are derived from unsaturated fatty acids (9). The precursors may also be added as such to the food product before processing, in which case the meat flavor is formed during cooking.

The basic meat flavor may also be derived from natural products such as meat extract, chicken powder or a smoke solution. In these cases, of course, the compounds present in the natural materials have performed as the flavor precursors. It goes without saying that the use of more traditional flavorings like salt and spices remains indispensable (6). The cases in which the tools i.e. the meat flavor mixtures are applied may be subdivided into three categories.

Processed Meat Products. First there is the processed meat product which during processing, e.g. heat sterilization or drying, has lost some of the flavor characteristics that are usually present in the equivalent freshly cooked food. In some cases artifacts i.e. flavor compounds not predominant in the fresh food may be formed. These are usually perceived as off-flavors. The aromagram or "skyline" will look like Figure 12.

Generally the most volatile or the most labile flavor compounds have disappeared. This is most readily noticed by the lack of "tall buildings". Changes in the smaller buildings of the skyline are less dramatic. The basic meat flavor is therefore still pre-

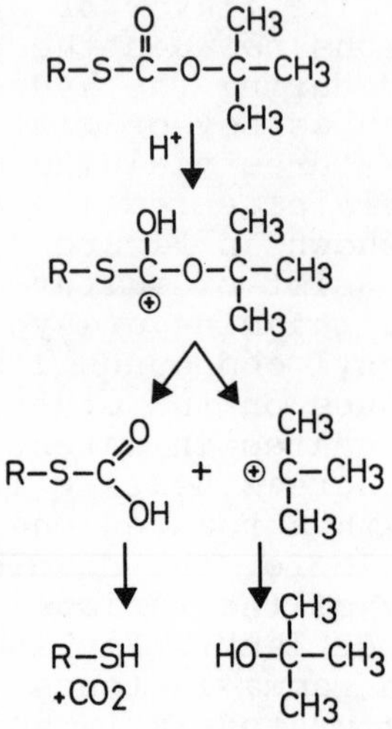

Figure 10. Use of O-T-alkyl thiocarbonates of thiols for flavor release

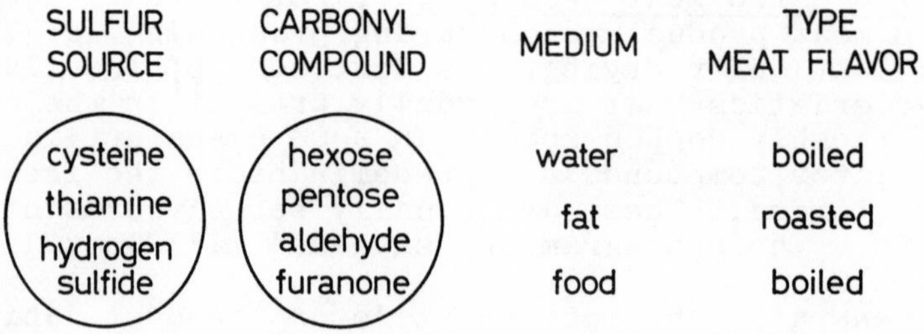

Figure 11. Preparation of meat flavors from precursors

sent. In addition, buildings that do not belong to
the skyline appear. These are "artifacts". Examples
of artifacts due to processing are excesses of aldehydes or of certain sulfur compounds like hydrogen
sulfide, methanethiol and dimethyl sulfide in canned
meat products. This effect may be prevented or at
least reduced by the choice of proper sterilization
conditions and of pH, or the addition of inhibitors
like salts of fumaric, maleic or sorbic acid (10).
Such processed foods, in which the basic meat flavor
is still present can usually best be flavored by the
addition of a mixture of top-notes, either in excess
or in a stabilized form.

Products with Little Meat. The second type of
application is in products that for economic reasons
part of the meat has been left out of the formulation.
This results in a dilution of the meat flavor. An example is a fabricated soup, which is not usually prepared in the same way as a home-made soup. The meat
flavor in such a product should be enhanced by taste
mixtures and mixtures based on precursors e.g. hydrolysates. As these products also fall in the first category, top-note mixtures are recommended as well.

Products without Meat - Vegetable Protein Products.
Finally, there is the extreme case of a meat-like
product which contains no meat at all. Examples are
some of the products based on textured vegetable protein materials, e.g. the so-called full analogues (8).
The development of such products, which is currently of
great interest, provides a major challenge to the
flavor chemist, and it therefore is worthwhile to
dwell upon it for a few moments.
    Textured vegetable protein materials are a new
type of food ingredients, which may appeal to the consumer provided the food itself fits into the general
consumption pattern. Since the nutritional properties
of these vegetable protein materials, particularly soy
protein materials, resemble those of meat, it is logi-
cal to allow meat to serve as a model for these new
ingredients. It should, however, be stressed that
vegetable protein materials should not be seen just as
meat imitations, but rather as novel, protein-rich
food ingredients, no more than spaghetti should be seen
as an imitation of potatoes.
    There are some problems associated with the fla-
voring of textured soy protein materials. Appearance
and texture of the products based on them are different
from those of existing products; this difference in

texture results in a different mode of flavor release
during chewing. Then there is the typical soy flavor
in the unrefined soy materials to which people do not
grow adjusted. Masking of this off-flavor, for instance with strong spicy and savory notes is not
satisfactory, as the off-flavor penetrates the masking
flavors on repeated consumption. The effect is not
unlike that of painting a car without removing the old
paint. After some weeks the new paint will peel off
and the old layer will gradually become visible. The
only way in which the incorporation of soy protein
materials at high levels in food products can be made
acceptable is by removing the off-flavor compounds and
their precursors, for instance by solvent extraction.
People in the Orient have always been aware of this,
since the traditional soy products consumed in those
countries have been pretreated in some way, e.g. by
extraction or fermentation, which effects the removal
of the off-flavors. If we assume that the off-flavor
inherent to the soy protein materials has been removed,
its aromagram will be similar to that shown in
Figure 13. Thus, the meat flavor of the product to be
developed has to be built up from scratch. This would
mean therefore a combination of non-volatile taste
compounds, basic meat flavor (the small buildings) and
the top-notes (tall buildings). The most difficult
part in this seems to be the basic meat flavor; the
combination of taste compounds and top-notes alone
leads to a synthetic, unreal flavor impression. Meat
flavors based on precursor systems therefore play a
very important role in this type of application. Of
course, it helps tremendously if some natural meat
flavor is present in the form of meat itself or as
meat extract, chicken powder, etc. The food product
then moves into the second category mentioned before,
in which only part of the natural meat ingredient has
been replaced, and the natural flavor is only diluted.
Therefore, textured vegetable protein materials will
frequently be introduced as meat extenders.

      In this context I would like to comment on the
recommendations of some legislative authorities concerning the supplementation of soy protein materials
with free methionine with the aim of enhancing their
protein value. Apart from the question whether this
is necessary in view of the reasonably high quality of
soy protein and the more than adequate protein intake
in Western countries, there are a number of disadvan-
tages associated with such supplentations, which are
relevant to the subject of this paper. Free methionine is labile and acts as a precursor of volatiles.

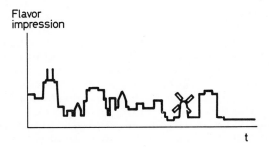

Figure 12. *Hypothetical aromagram of processed meat flavor*

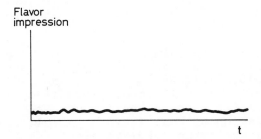

Figure 13. *Aromagram of flavorless meat analog*

It has been found that 30 - 80% of the added methionine is degraded, depending on the type of processing (11). Thus, only part of the added methionine is effective as a nutritional supplement; the rest gives rise to artifacts, some of which, e.g. methional, cause objectionable off-flavors. And the levels of methionine recommended for nutritional fortification are an order of magnitude higher than the levels of amino acids normally used for flavoring purposes.

Conclusion. In this survey I have attempted to indicate the importance of acquiring knowledge on the role of precursors in the formation of flavor. At present it is still very difficult to compound complicated flavors like meat flavor without using some of the natural product itself, or an extract thereof as the foundation onto which the added flavor is built. Appropriate use of precursors may lead to a wide range of processed and novel food products, which partly on account of their good flavor characteristics will be acceptable and even attractive to the consumer in our sophisticated Western society. This may be a goal in itself, but it is also the only way by which increased amounts of food can be made available to the growing world population.

ABSTRACT

Three types of flavors may be distinguished: taste mixtures composed of non-volatiles; mixtures of volatile compounds ("top-notes"); like furanones, and so-called "reaction flavors". The latter comprise precursors which react upon heating to yield mixtures with a basic flavor balancing the rather harsh top-notes.
The precursors of meat flavor generally consist of sulfur compounds like cysteine, thiamine or hydrogen sulfide releasers and carbonyl compounds like monosaccharides, aldehydes or furanones. They may react in water, fat or the food product itself. Each of the three flavor types makes its specific contribution to the formation of meat flavor. A well balanced combination of the three leads to a total meat flavor, although the presence of some natural meat flavor is still required for best results.

Literature Cited
1. Kramer, A., Food Technol., (1972) 26, 34.
2. Hodge, J.E., Mills, F.D. and Fisher, B.E., Cereal Sci. Today (1972) 17, 34.

3.   Schutte, L., C.R.C. Crit. Rev. Food Technol., <u>4</u>
     (1974) <u>4</u>, 457.
4.   Rizzi, G.P., J. Agric. Food Chem., (1972) <u>10</u>,
     1081.
5.   Sulser, H., DePizzol, J. and Büchi, W., J. Food
     Sci., (1967) <u>32</u>, 611.
6.   Blake, A., Food Manuf., (1975) <u>50</u>, 35.
7.   Balassa, L.L. and Fauger, G.O., C.R.C. Crit. Rev.
     Food Technol., (1971) <u>1</u>, 245.
8.   Wilson, R.A. and Katz, I., The Flavor Industry,
     (1974) <u>5</u>, 30.
9.   Harkes, P.D. and Begeman, W.J., J. Am. Oil Chem.
     Soc., (1974) <u>51</u>, 356.
10. Persson, T. and Von Sydow, E., J. Food Sci.,
     (1974) <u>39</u>, 406.
11. Shemer, M. and Perkins, E.G., J. Agric. Food
     Chem., (1975) <u>23</u>, 201.

7

Reaction Products of α–Dicarbonyl Compounds, Aldehydes, Hydrogen Sulfide, and Ammonia

HENK J. TAKKEN, LEENDERT M. VAN DER LINDE, PIETER J. DE VALOIS, HANS M. VAN DORT, and MANS BOELENS

Naarden International Research Department, P.O. Box 2, Naarden-Bussum, The Netherlands

Sulfur compounds have been found in many aromas as well as in several essential oils. Although the concentration of these compounds is often very low the contribution to the overall flavor can be important due to their odor strength. As a consequence of the low concentration and in some cases the lack of stability the analysis can be extremely difficult. Therefore much attention is paid to the investigation of model systems in which only a few compounds are present which are considered to be precursors of flavor components.

An extensive survey of sulfur compounds detected in model systems has been given by Schutte in his review on precursors of sulfur containing flavor compounds (1).

Reaction products of aldehydes and hydrogensulfide

Our approach was to investigate the compounds formed by reaction of important flavor components with hydrogensulfide, ammonia and thiols. We first investigated the reaction products of aldehydes and hydrogensulfide. At atmospheric pressure, mainly cyclic trimers like dioxathianes, oxadithianes and trithianes were formed. However, in a closed glass vessel with excess of hydrogensulfide the reaction mixture consisted mainly of 1,1-alkanedithiols and bis-(1-mercaptoalkyl)sulfides (2). Although these compounds have not been found in aromas they can be intermediates for a number of important flavor compounds. The following reactions with bis-(mercaptoalkyl)sulfides have been observed (figure 1):
- under oxydative circumstances dialkyltrithiolanes are readily formed
- under the influence of acids a complete conversion into trialkyltrithianes does occur
- at elevated temperatures isomerisation into trisulfides occurs, which compounds disproportionate into di- and tetrasulfides
- with ammonia dithiazines are formed.

114

Reaction products of ∝-dicarbonylcompounds, aldehydes, hydrogensulfide and ammonia

We now wish to report on our results of the analysis of reaction mixtures obtained from ∝-dicarbonylcompounds, aldehydes,
hydrogensulfide and ammonia. The following combinations of
∝-dicarbonylcompounds and aldehydes were used:
2,3-butanedione + acetaldehyde
2,3-butanedione + propionaldehyde
2,3-butanedione + crotonaldehyde
2,3-pentanedione + acetaldehyde
2-oxopropanal + acetaldehyde
The experiments were carried out in closed glass vessels in
buffered solutions (pH 5) at 20° and 90°C.
The reactants were used in the following molar ratio:
dicarbonylcompound: aldehyde : hydrogensulfide : ammonia =
1 : 1 : 2 : 1. The ether extracts of the reaction mixtures were
analysed with a mass spectrometer directly coupled to a gas-
chromatograph.

In all reaction mixtures, with the exception of that of
crotonaldehyde, we observed the presence of 1,1-alkanedithiols and
3,5-dialkyl-1,2,4-trithiolanes, originating from the aldehydes.
Starting from butanedione 2-mercaptobutanone-3 was formed (figure
2), while in the mixtures derived from pentanedione both possible
mercaptopentanones were found. In a few cases the formation of
3,6-dimethyl-1,2,4,5-tetrathiane was observed. The diethyl analog
of this compound was found by Ledl (3) in a reaction mixture of
propionaldehyde, hydrogensulfide and ammonia. They can be regarded
as oxidation products of the initially formed alkanedithiols.
Two other nitrogen free compounds have been found, namely 2-
mercapto-butene-2 and 3,5,6-trimethyl-1,2,4-trithiane (figure 3).
The first product can be derived from mercaptobutanone, the latter
can be formed from the intermediate butanedithiol, acetaldehyde
and hydrogensulfide. In the reaction mixture of butanedione with
propionaldehyde a compound derived from the intermediately formed
butenedithiol was detected namely 2-ethyl-4,5-dimethyl-1,3-di-
thiolene. The corresponding compound derived from 2-oxopropanal
was also detected. So, although they have never been found, 2,4,5-
trimethyl-1,3-dithiolane and 2,4,5-trimethyl-1,3-dithiolene can
be expected to be flavor components.
The compounds mentioned so far can be formed without ammonia.
We will now pay attention to the nitrogen containing products. In
all reaction mixtures, again except that of crotonaldehyde, the
2,4,6-trialkyldihydro-1,3,5-dithiazines were found. These compounds are readily formed from aldehydes, hydrogensulfide and
ammonia. The trimethylderivative, known as thialdine, is identified by Brinkman (4) in beef broth and by Wilson (5) in boiled
meat.

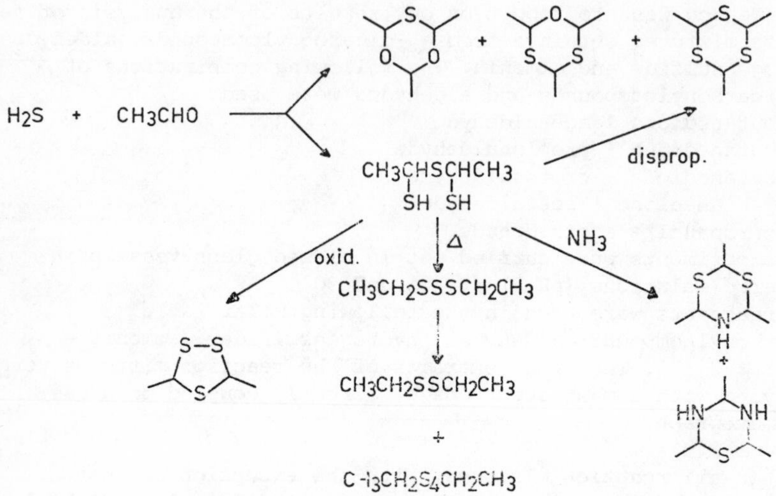

Figure 1. Reaction products of acetaldehyde and hydrogen sulfide

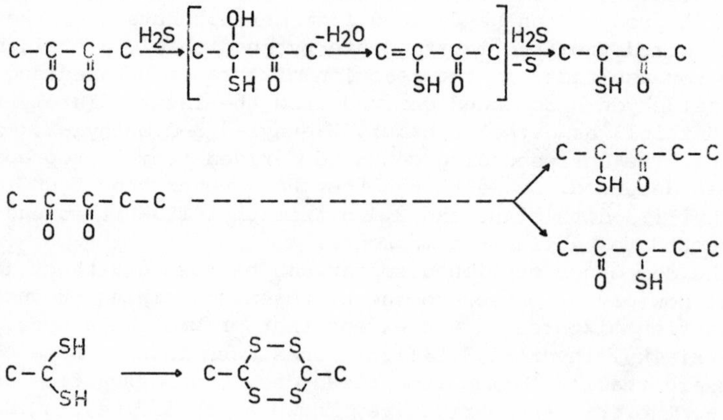

Figure 2. Formation of α-mercaptoketones and dimethyltetrathiane

The concentration of the dithiazine formed is always lower in the high temperature experiment. Thialdine seems to be rather unstable under the reaction conditions chosen. Ledl (6) recently published the results of his investigation of the aroma of roasted onion. He was surprised not to be able to detect trialkyldihydrodithiazines. Viewed in the light of the above mentioned unstability this is not surprising because he carried out his experiments at 150°C.

The most striking difference between the low and high temperature experiments is the formation of two isomeric hydroxythiazolines at room temperature (figure 4). These compounds are absent in the reaction mixtures of 90° experiments, probably due to a dehydration reaction into thiazoles. Besides these thiazoles thiazolines are present.

The combination of butanedione and propionaldehyde yielded the expected thiazoles and thiazolines, but also the trimethyl derivatives were detected. A possible reaction route leading to these products is given in figure 5. We observed the formation of 2-butanone and its incorporation in a thiazoline derivative.

Due to the asymmetric structure of 2,3-pentanedione the formation of two sets of thiazoles, thiazolines and hydroxythiazolines can be expected and indeed proved to be present (figure 6).

After reaction of 2-oxopropanal, acetaldehyde, hydrogensulfide and ammonia the expected dimethylthiazoles and thiazolines were detected (figure 7). Moreover, 2-acetyl-4(and 5) -methylthiazole are formed due to a reaction of 2 moles pyruvaldehyde. 2-Acetyl-4-methylthiazole is also found by Kato et al. (7) in reaction mixtures of cystine and pyruvaldehyde.

Besides the mentioned thiazoles the corresponding oxazoles were present in all reaction mixtures.

Trimethylisothiazole was detected in the mixtures obtained after reaction of butanedione, acetaldehyde, hydrogensulfide and ammonia, preferably at low temperature. We have synthesized a number of these isothiazoles from 1,3-diketones with hydrogensulfide, ammonia and sulfur (figure 8). It seems likely that starting from 2-mercaptobutanone via aldolcondensation with acetaldehyde a 1,3-dione is formed in our experiments. Transformation into isothiazole may than occur according to the route given in figure 8.

We observed a large similarity in the composition of the reaction mixtures from the different combinations of dicarbonylcompounds and aldehydes, however, with the exception of the mixture obtained from butanedione and crotonaldehyde with hydrogensulfide and ammonia. The main product in this reaction mixture was 3-mercaptobutanal. Moreover we detected 3-methyl-1,2-dithiolane and 3-methyl-1,2-dithiolene. In figure 9 a route for the formation of these compounds is suggested. In the same mixture we observed the presence of 5,6-dihydro-2,6-dimethylthiopyrane-2H-3-carboxaldehyde and 2,6-dimethyl-2H-thiopyrane-3-carboxaldehyde. The first aldehyde is also found by Badings et al. (8) in reaction mixtures of crotonaldehyde and hydrogensulfide.

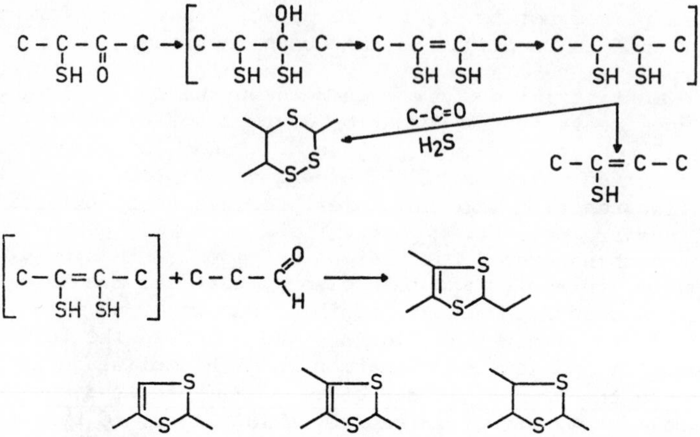

Figure 3.   Formation of 2-mercaptobutene-2, 1,2,4-trithiane, and 1,3-dithiolenes

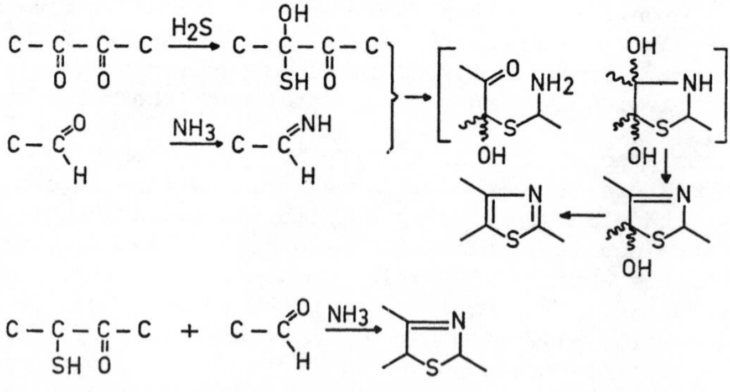

Figure 4.   Trimethyl-hydroxythiazolines and trimethylthiazole

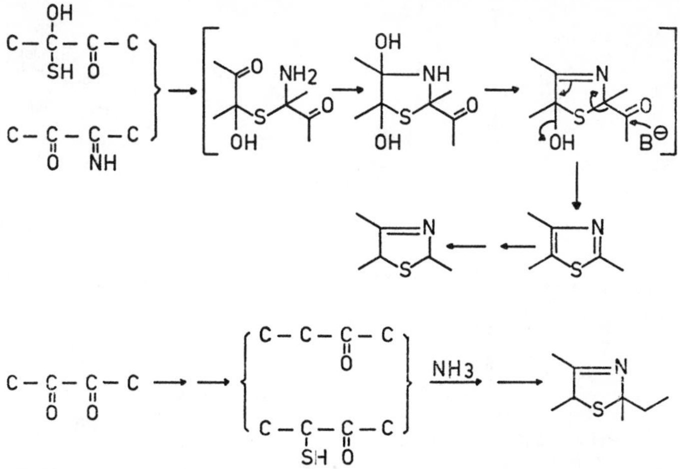

Figure 5. Trimethylthiazole, trimethylthiazoline, and ethyltrimethyl-thiazoline from butanedione, H₂S, and NH₃

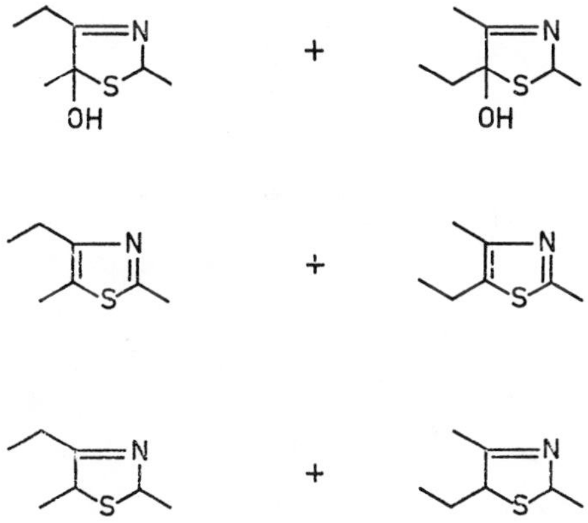

Figure 6. Hydroxythiazolines, thiazoles, and thiazolines from 2,3-pentanedione, acetaldehyde, H₂S, and NH₃

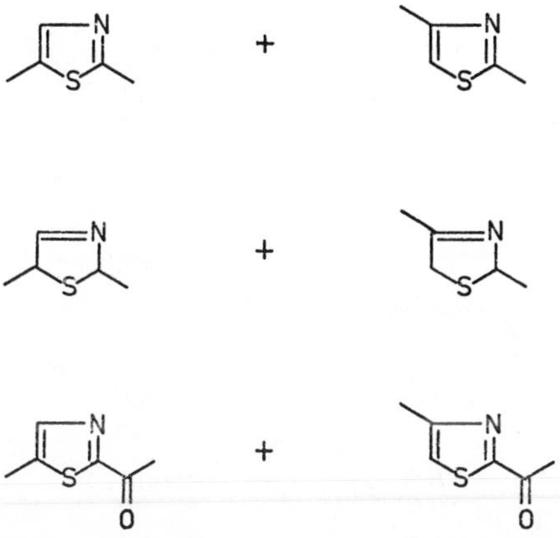

Figure 7. Thiazoles, thiazolines, and acetylthiazoles from pyruvaldehyde, acetaldehyde, H₂S, and NH₃

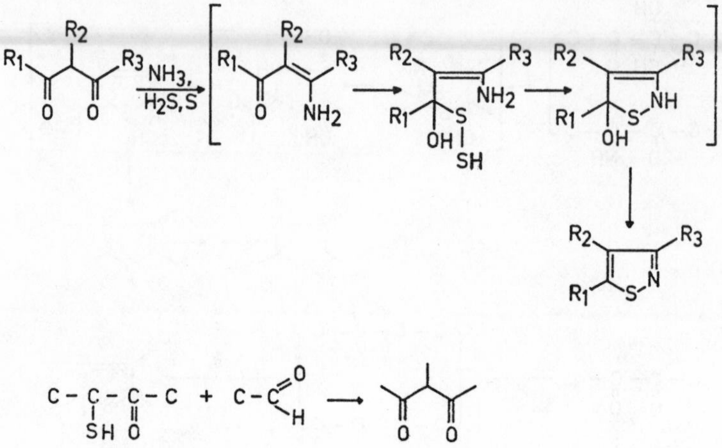

Figure 8. Formation of isothiazoles

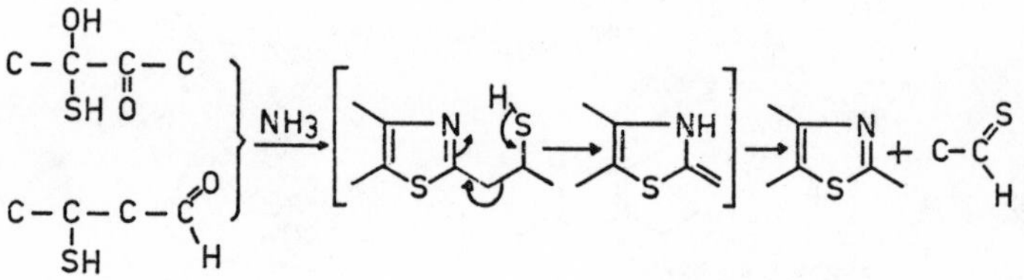

Figure 9. Reaction products from crotonaldehyde and H_2S

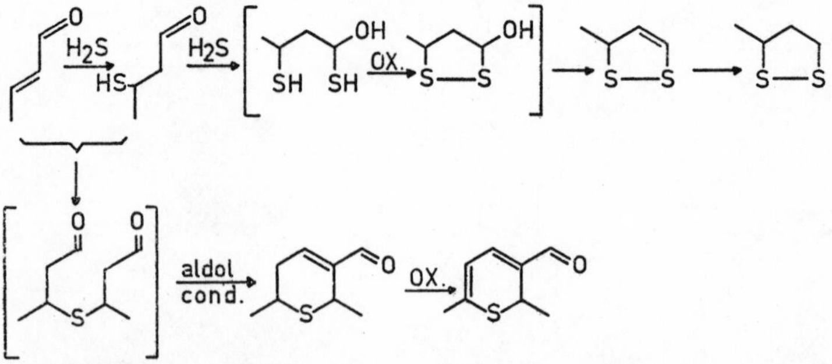

Figure 10. Formation of trimethylthiazole from butanedione, crotonaldehyde, H_2S, and NH_3

Reactions with crotonaldehyde yielded only trimethyloxazole and thiazole. No products with a propenyl or mercaptopropyl group were present. This might be due to a quick retro aldol reaction of the initially formed mercaptopropylthiazole (figure 10).

Conclusion

The results of our analysis of reaction products from α-di-carbonyl compounds, aldehydes, hydrogensulfide and ammonia show that a number of already known flavor components are formed. Moreover several so far unknown compounds have been detected. These compounds can be expected in aromas. Therefore we believe that study of this type of model systems can be of great help to the analysis of flavor complexes.

Literature cited

1. Schutte, L., Critical Reviews in Food Techn., March 1974, 457.
2. Boelens, H., van der Linde, L.M., de Valois, P.J., van Dort, J.M., Takken, H.J., J.Agr.Food Chem. 22, 1071 (1974).
3. Ledl, F., Z.Lebensm.Unters.-Forsch. 157, 28 (1975).
4. Brinkman, H.W., Copier, H., de Leeuw, J.J.M., Tjan, S.B., J.Agr.Food Chem. 20, 177 (1972).
5. Wilson, R.A., Mussinan, C.J., Katz, I., Sanderson, A., J.Agr.Food Chem. 21, 873 (1973).
6. Ledl, F., Z.Lebensm.Unters.-Forsch. 157, 229 (1975).
7. Kato, S., Kurato, T., Fujimaki, M., Agric.Biol.Chem. 37, 539 (1973).
8. Badings, H.T., Maarse, H., Kleipool, T.J.C., Tas, A.C., Naeter, R., ten Noever de Brauw, M.C., CIVO Aroma Symposium, May 1975. Zeist. The Netherlands.

8

Non-Enzymic Transamination of Unsaturated Carbonyls: A General Source of Nitrogenous Flavor Compounds in Foods

GEORGE P. RIZZI

The Procter & Gamble Co., Winton Hill Technical Center, Cincinnati, Ohio 45224

Volatile nitrogen compounds have long been recognized as significant contributors to food flavors. The compounds range in complexity from the simplest alkylamines commonly associated with the aroma of fish to diverse heterocyclics such as pyridines, pyrroles and pyrazines which form during the cooking or roasting of foodstuffs.

α–Amino acids are frequently implicated as precursors to volatile flavor components and reaction pathways have been described to explain the transfer of nitrogen to non-nitrogenous substrates. In this paper we present data which extends the scope of the Strecker degradation to allow for the incorporation of amino acid nitrogen into unsaturated aldehydes and ketones.

Perhaps the most thoroughly studied transamination reaction is that of pyridoxal (I), which through enzyme intervention is converted to the reductively aminated product pyridoxylamine (II) (1).

I	II	III

Except for its ring nitrogen compound I can be included in the general class of compounds, III, predicted (2) to be active participants in the Strecker degradation. In view of the reactivity of I, it is apparent that many types of conjugated carbonyl compounds like III might participate in Strecker-like reactions whether or not they contain a second carbonyl oxygen. More fundamentally one can predict, on the basis of the accepted mechanism (3),

that any compound which contains a carbonyl group attached
to a system of atoms capable of delocalizing a formal
negative charge, i.e. extended conjugation, should yield
Strecker-like reactions. The predicted net effect is
conversion of conjugated carbonyl substrates, generalized
in Figure 1, to their reductively aminated counterparts
with concomitant formation of the usual Strecker carbonyls.
This process differs from the usual, reductone mediated,
Strecker degradation in that a stable, monofunctional
product, i.e. an unsaturated amine, is obtained. In
contrast, the usual nitrogen-containing product is an α-
aminoketone which polymerizes or dimerizes to form pyrazines
(4).
    A wide variety of unsaturated, conjugated carbonyls
exist in foods, e.g. terpenes, fat autoxidation products
and flavonoids which if aminated during cooking could
become organoleptically more important. In this study our
main concern was with types of molecules found in foods
and models for study were selected on that basis.
    The reactions of amino acids and unsaturated, aliphatic
carbonyl compounds have received little attention according
to published literature. Similar reactions involving
aromatic carbonyls like benzaldehydes, acetophenones and
furfural occur readily and have been well documented by
Rizzi (5), (6) and Chatelus (7). Anet (8) has shown that
certain unsaturated aldose sugars react with amino acids
to form melanoidins, but specific nitrogeneous products
were not mentioned. Reactions of simple α, β-unsaturated
aldehydes and ketones were studied by Burton et. al. (9)
again with the emphasis on pigment formation. Unsaturated, aliphatic amines, have not been reported in foods,
however, analogous saturated alkyl amines have been recognized
and they are believed to be formed during thermal and/or
enzymic breakdown of amino acids and phosphatides.

Experimental Section

    Model systems were examined to test our theoretical
predictions. In general, equimolar quantities of dl-β-
phenylalanine (PA) or dl-2-amino-2-methylbutyric acid
(AMBA) and various unsaturated carbonyl substrates were
heated in refluxing diglyme for 1-2 hours. Under these
conditions decarboxylation occurred readily and one-phase
reaction mixtures were usually obtained by the end of the
reaction period. Diglyme was removed by vacuum distillation
and residues were hydrolyzed with N HCl at 25°C to convert
imines completely to amines before analysis. The hydrolysates
were extracted with ether to separate neutral and acidic
products and samples of the aqueous-HCl phase were analyzed
by either horizontal paper chromatography (HPC) using

Whatman No. 1 paper and the upper phase of an n-butanol-
acetic acid-water (4:1:5 v/v/v) mixture for elution or by
thin layer chromatography on Silica Gel - G plates and a
4:1:1 ratio of the same solvents. In all cases amines
were visualized by spraying with dilute ninhydrin solution
and heating briefly in an oven at 80-100°C. In some
cases, individual pure products were separated by distillation
or crystallization and identified by customary physical
and chemical means (cf. examples below).

Reference compounds were prepared by standard methods
of organic synthesis. Infrared (IR) spectra were recorded
on a Perkin Elmer Infracord spectrophotometer and melting
points were taken in open capillaries and are uncorrected.
Ultraviolet (UV) spectra were taken with a Cary Model 14
instrument. Preparative GLC separations were performed
isothermally with an Aerograph A-90-P unit using a 15 ft.
x 0.25 in. stainless steel column containing 15% Carbowax
20M on 60/80 mesh Chromosorb W at 200°C. Amino acids and
organic reagents were obtained from the Aldrich Chemical
Company, Milwaukee, Wisconsin 53233, or Matheson, Coleman
& Bell, Norwood, Ohio 45212.

Reaction of AMBA with Citral.

A slurry of AMBA (3.36 g, 0.0286 mol), 4.50 ml
(0.0263 mol) of citral and 30 ml of diglyme was stirred
and refluxed under a N_2 atmosphere 2 hrs. The mixture was
hydrolyzed with 30 ml N HCl (3 hrs.) and neutral products
were separated by ether extraction (3 x 100 ml). Evaporation
of the aqueous phase gave crude geranylamine hydrochloride
which was treated with Na_2CO_3 to liberate the free base.
Ether extraction followed by distillation gave 1.220 g
(30.3% yield) of geranylamine, bp 100-102°C/13 mm, n_D^{24} 1.4711.
In a separate experiment the amine hydrochloride was
recrystallized from isopropanol to give 25% yield of
colorless product, mp 145.5 - 146°C, literature reported
mp 145-146°C, (10). Calcd for $C_{10}H_{20}$ NCl: C, 63.4; H,
11.55; N, 7.39. Found: C, 63.5; H, 11.1; N, 7.3.

Reaction of PA with d-Carvone.

PA (3.727 g, 0.0226 mol), d-carvone (3.12 ml, 0.020
mol) and 30 ml of diglyme were refluxed 1.25 hr. under
nitrogen in which time all solids originally present had
disappeared. Hydrolysis was effected by adding 30 ml of N
HCl and refluxing 1 hr. The cooled mixture was diluted
with water (3 volumes) and extracted with ether to provide
2.646 g (88%) recovered d-carvone, authentic by IR analysis.
Concentration of the aqueous phase gave 3.393 g (96%
yield) of 2-phenylethylamine hydrochloride. Recrystalli-

zation from isopropanol gave 2.737 g, mp 218-223°C; the IR
spectrum was identical with a spectrum of the authentic
material.

Reaction of AMBA with <u>Trans-2-Hexenal</u>.

<u>Trans-2-hexenal</u> (2.31 ml, 0.020 mol) was added to a
stirred slurry of AMBA (2.34 g, 0.020 mol) in 30 ml of
freshly redistilled diglyme (bp 59°C/12 mm). The mixture
was refluxed under nitrogen 2 hrs, cooled, diluted with
ether and filtered to yield 0.432 g of unchanged AMBA
(18%). The filtrate was evaporated to remove ether and 20
ml of N HCl was added to effect hydrolysis. After 2 hrs
at 25°C, the mixture was diluted with water (100 ml),
extracted with ether (3 x 100 ml) and the aqueous-diglyme
phase was concentrated under vacuum to yield 4.28 g of
moist crystals. HPC analysis indicated a single ninhydrin
positive product, R_f 0.71, yellow-tan color changing to
purple on standing. The crude product was washed with
ether and recrystallized from ethyl acetate to give 0.84 g
of tan plates, mp 194-209°C dec. (38% yield). Final
recrystallization from ethanol/ethyl acetate gave analytically
pure <u>trans-2-hexenylamine</u> hydrochloride as colorless,
broad plates, mp 207-210°C, IR (**KBr** pellet) 10.25u (<u>trans</u>
C = C, no <u>cis</u> absorption observed). Calcd for: C_6H_{14}
NCl: C, 53.2; H, 10.3; N, 10.3; Cl 26.2. Found: C, 53.1;
H, 10.6; N, 10.0; Cl, 26.8. The free amine obtained by
basifying the hydrochloride with sodium hydroxide was an
oil, bp 33-37°C/12 mm, n_D^{25} 1.4373.

2,4 Hexadienylamine.

Hydroxylamine hydrochloride (3.544 g, 0.051 mol) was
added to a mixture of 2,4-hexadienal (5.29 ml, 0.050 mol)
and 3N NaOH (50 ml) and stirred for 2 hrs at ambient
temperature. Neutralization at 0°C with 4N HCl followed
by ether extraction (5 x 100 ml), drying of the ether
(MgSO$_4$) and concentration gave crude 2,4-hexadienaldoxime,
<u>ca.</u> 5 g. The oxime was dissolved in a mixture of ethanol
(80 ml) and acetic acid (80 ml) and cooled to 0°C. While
stirring at 0° 40 g of powdered zinc was added gradually
during 1 hr. The reaction was completed by heating at
40°C for 0.5 hrs and the solids were removed by filtration.
The filtrate was concentrated to remove ethanol, basified
to pH 12 with 33% aqueous KOH solution and extracted with
ether to provide 4.16 g of crude amine. Distillation gave
1.5 g of 2,4-hexadienylamine, bp 51-55°C/12 mm, n_D 1.4970.
The amine was characterized as its N-acetyl derivative
prepared in pyridine with acetic anhydride. After recrystallization from benzene/hexane the acetamide was obtained as

colorless plates, mp 82.5-83.5°C; UV $_{max}^{\varepsilon\tau OH}$ 228 nm (log ε 4.51); IR (CHCl$_3$) 2.92 u (NH), 5.98 u (amide C=O), 10.1 u (trans C=C). Calcd for C$_8$H$_{13}$NO: C, 69.1; H, 9.35; N 10.07. Found: C, 69.4; H, 9.3; N 9.9.

Trans-Cinnamylamine.

The two-step procedure of Gensler and Rockett (11) was used without essential modification except that cinnamyl bromide was substituted for the chloride because of its availability. N-cinnamylphthalimide was obtained in 73% yield, mp 152-154°C (reported, mp 153-153.5°C). Following hydrazinolysis cinnamylamine hydrochloride was isolated as glistening, colorless plates, mp 225-230°C (reported 209-219°C). Calcd for C$_9$H$_{12}$NCl: C, 63.8; H, 7.08; N, 8.27. Found: C, 64.1; H, 7.4; N, 8.2.

3-Amino-1-phenyl-1-butene (V)

The oxime of trans-benzalacetone was reduced with zinc dust in acetic acid (12). The unsaturated amine was characterized as its N-benzoyl derivative, mp 134-137°C, (VI). Baddiley reported mp 137°C (12).

Dihydrobenzalacetone.

Benzalacetone (14.62 g, 0.10 mol) in 100 ml of ethyl acetate was treated with 0.1 g of 10% palladium-on-charcoal catalyst and hydrogenated in a Parr apparatus. The hydrogen pressure (initially 42.0 lb/in^2) fell to 33.5 lb/in^2 during 90 min which corresponded to an uptake of one mol equiv of H$_2$. Following filtration of catalyst and removal of solvent the residual oil was distilled to afford 13.6 g of dihydrobenzalacetone, bp 115-117° C/12 mm; IR (film) 5.80 u (saturated C=O).

Results

A. Reactions of Aldehydes.

Trans-2-hexenal, a well known constituent of food flavors (especially off flavors resulting from fat autoxidation), reacted vigorously with PA in refluxing diglyme. Paper chromatographic analysis (HPC) of the product after HCl hydrolysis revealed two major ninhydrin positive materials. The major product (purple spot, 60% relative yield) was shown to be 2-phenylethylamine by comparison of its R$_f$ with the R$_f$ of an authentic sample. A second product (tan spot, 40%), obviously an amine was observed at an R$_f$ somewhat greater than 2-phenylethylamine. Reaction

of trans-2-hexenal with AMBA led to a preponderance of the
second product which was isolated and proved to be trans-
2-hexenylamine.

Citral, a natural mixture of the stereoisomeric
terpenes geranial and neral, reacted similarly with PA.
Besides 2-phenylethylamine (70% relative yield) a second,
less polar substance was observed (yellow spot with
ninhydrin, 30%) which was subsequently identified as
geranylamine, the predicted product of reductive amination.
Upon reaction of citral with AMBA, geranylamine proved to
be the major reaction product thus facilitating its isolation
and characterization.

2,4-hexadienal reacted too vigorously with both amino
acids at 160°C and milder reaction conditions, 120°C/1 hr,
had to be employed. In addition to 2-phenylethylamine
(70%), a second, light yellow spot appeared at higher R_f
(30%). Reaction of the dienal was slower with AMBA as
evidenced by more recovered amino acid but the major
product isolated was shown to be trans, trans-2,4-
hexadienylamine, the product of reductive amination.

Trans-cinnamaldehyde reacted readily with PA in
boiling diglyme; and, after hydrolysis and concentration
of the HCl solution, 2-phenylethylamine hydrochloride
separated in crystalline form, mp 210-222°C dec. (after
recrystallization from isopropanol). The infrared spectrum
of the reaction product was identical with a spectrum of
the authentic material. HPC analysis of the mother liquor
indicated besides additional 2-phenylethylamine at R_f 0.70
an overlapping, but easily distinguishable, yellow spot at
R_f 0.74. Subsequent reaction of trans-cinnamaldehyde with
AMBA led to a preponderance of the second substance which
was isolated and shown to be trans-cinnamylamine.

B. Ketones.

Conjugated, aliphatic ketones do not occur as abundantly
in foods as unsaturated aldehydes. d-Carvone, a widespread
terpene ketone, reacted completely with one equivalent of
PA in boiling diglyme after 1.25 hours. Hydrolysis of the
reaction mixture gave a 96% yield of crystalline 2-phenyl-
ethylamine hydrochloride; 67% yield after recrystallization,
mp 218-223° dec. No evidence was found for d-carvylamine,
the expected reductive amination product. Reactions of d-
carvone and AMBA proved to be extremely slow at ca. 160°C;
however, decarboxylation of the AMBA did take place to a
slight extent as evidenced by formation of a precipitate
when gaseous reaction products were swept into a 5% $Ba(OH)_2$
solution with N_2. After two hours most of the AMBA was
recovered unchanged and the neutral products isolated
after HCl hydrolysis contained 43% recovered d-carvone

plus a 51% yield of cavracol (2-methyl-5-isopropylphenol).
The latter material apparently resulted via acid catalyzed
isomerization of d-carvone.

Because of the apparent low reactivity of d-carvone
with AMBA a ketone containing more extensive conjugation,
i.e. trans-benzalacetone (IV) was tried. Reaction of IV
and PA gave, after HCl hydrolysis, a quantitative yield of
2-phenylethylamine similar to the d-carvone reaction. No
trace of the expected reductive amination product, 3-
amino-1-phenyl-1-butene (V) was detected by HPC. Reaction
of IV and AMBA led to 39% consumption of the amino acid in
3 hours. Hydrolysis and the usual workup gave the amine V,

$$C_6H_5-\overset{\text{H}}{\underset{\text{H}}{C}}=C-\underset{\overset{\|}{O}}{C}-CH_3 \qquad C_6H_5-\overset{\text{H}}{\underset{\text{H}}{C}}=C-\overset{\text{H}}{\underset{\text{NHR}}{C}}-CH_3$$

$$\text{IV} \qquad\qquad \begin{array}{l}\text{V, } R=H \\ \text{VI, } R=\underset{O}{C}C_6H_5 \end{array}$$

HPC R_f 0.79, in 48% yield (based on amino acid consumed)
and a trace of sec.-butylamine, R_f 0.68. The neutral
products of the reaction consisted of recovered IV 90.8%
(80% recovery) and dihydrobenzalacetone 9.2% (8% yield).
The latter product was isolated by preparative gas chromatography (GC) and its structure was established by comparing
GC retention times (R_t) and infrared spectra with authentic
data.

Discussion

When α, β-unsaturated aldehydes and ketones were
heated with β-phenylalanine (PA) or 2-amino-2-
methylbutyric acid (AMBA) and the initial reaction products
were hydrolyzed, unsaturated amines representing products
of carbonyl reductive amination were obtained (Table I).
Other products formed included CO_2, saturated amines
derived from the amino acids and saturated carbonyls.
Generally PA (an α-monosubstituted amino acid) yielded 2-
phenylethylamine, the product of simple decarboxylation
plus unsaturated amines, while AMBA (an α-disubstituted
amino acid) usually gave unsaturated amines exclusively.
Qualitatively, the rate of reaction increased with the
degree of extended conjugation, i.e. for aldehydes:
dienals > enals > saturated carbonyls. This is consistent
with our hypothesis which predicts faster reactions in
systems capable of extensive charge delocalization.

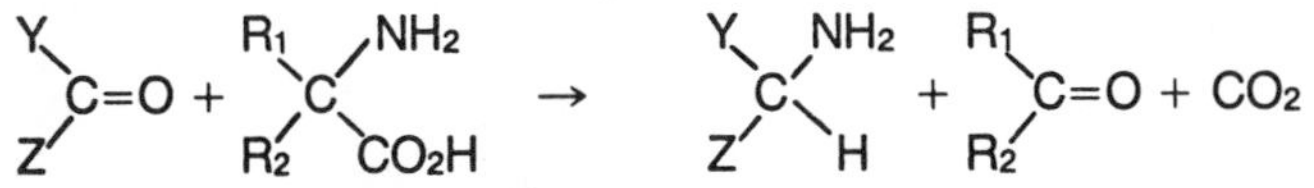

$$\begin{array}{c}Y\\Z\end{array}C{=}O \;+\; \begin{array}{c}R_1\\R_2\end{array}C\begin{array}{c}NH_2\\CO_2H\end{array} \;\rightarrow\; \begin{array}{c}Y\\Z\end{array}C\begin{array}{c}NH_2\\H\end{array} \;+\; \begin{array}{c}R_1\\R_2\end{array}C{=}O \;+\; CO_2$$

Y, Z = H, Alkyl Group or conjugated elements of unsaturation.

R_1, R_2 = H or Alkyl Group

Figure 1. Generalized reductive amination scheme

Table I

Products Identified in Reactions of α-Amino Acids and α,β-Unsaturated Carbonyl Compounds

Aldehydes	Amino Acid[a]	Unsat. Amines	Sat. Amines	Sat. Carbonyls
trans-2-Hexenal	PA	2-Hexenylamine	2-phenylethylamine	—
	AMBA	2-Hexenylamine (31)[b]	—	—
Citral	PA	Geranylamine	2-phenylethylamine	—
	AMBA	Geranylamine (30)		
trans-Cinnamaldehyde	PA	Cinnamylamine	2-phenylethylamine	Dihydrocinnamaldehyde[c]
	AMBA	trans-Cinnamylamine (30)		
2,4-Hexadienal	PA	2,4-Hexadienylamine	2-phenylethylamine	—
	AMBA	2,4-Hexadienylamine	sec.-Butylamine	—
Ketones				
d-Carvone	PA	—	2-phenylethylamine (96)	Recovered d-Carvone
	AMBA	—	sec.-Butylamine	Recovered d-Carvone
trans-Benzalacetone	PA	—	2-phenylethylamine (100)	—
	AMBA	3-Amino-1-phenyl-1-butene (48)	sec.-Butylamine	Dihydrobenzalacetone (8)

[a] PA = d1-β-phenylalanine, AMBA = d1-2-amino-2-methylbutyric acid

[b] Numbers in parentheses represent % of theoretical yields based on the limiting reagent used.

[c] Trace product, tentatively identified by gas chromatography.

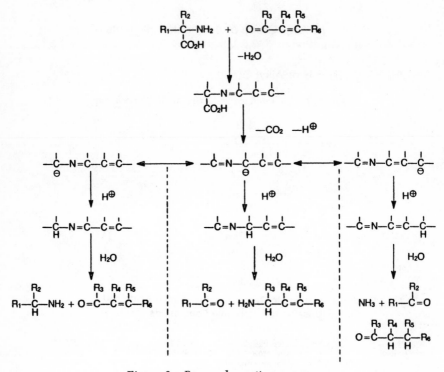

Figure 2. Proposed reaction sequence

Ketones showed less tendency to react than aldehydes perhaps due to steric factors involved in the initial carbonyl-amine addition step. Similar rate effects are known for reactions of simple amines with aldehydes vs. ketones (13). Qualitatively the disubstituted amino acid AMBA reacted more slowly with aldehydes and ketones than PA. This effect has been observed previously in the case of the Strecker degradation where the difference was attributed to inhibition of the carbonyl-amine addition by steric hindrance, cf. Friedman and Sigel (14).

A mechanism for reductive amination which accommodates our data is shown in Figure 2. Initial reaction of the amino acid and carbonyl proceeds with dehydration and loss of carbon dioxide to form a resonance stabilized azallylic carbanion whose sites of highest charge density are located at the (amino acid) α-carbon atom and on odd numbered carbons along the conjugated carbon chain. Subsequent protonation can lead to a mixture of imines and enamines which on hydrolysis gives the observed products. The exact nature and ratios of final products is a complex function of reaction conditions and the choice of substituents on the reactants. From our study it was clear that amino acids with two alkyl substituents led to a greater preponderance of reductive amination. This fact is consistent with our proposed mechanism since it is known that alkyl groups tend to destabilize neighboring carbanions (relative to hydrogen). This effect could retard protonation at the (amino acid) α-carbon atom and favor increased charge density elsewhere in the molecule. Under these circumstances protonation is more likely to occur at sites along the conjugated carbon chain.

When protonation occurs at the next alternate site adjacent to the (amino acid) α-carbon atom products of reductive amination will result. Protonation further down the conjugated chain is apparently rare but the fact it does occur, e.g. trans-benzalacetone conversion to dihydro-benzalacetone (Table 1), substantiates the proposed mechanism. No evidence was found for analogous dihydro products in other conjugated carbonyl reactions. Such a transformation, if it were more general, could be of great importance organoleptically, e.g. conversions of 2-hexenal to hexanal, 2,4-hexadienal to 4-hexenal, etc.

Literature Cited

1.   Kalyankar, G.D., Snell, Esmond E., Biochemistry, (1962), 1, 594.
2.   Schonberg, Alexander S., Moubasher, Radwan M., Mostafa, Akila M., J. Chem. Soc., (1948), 176.
3.   Baddar, Fawzy G., J. Chem. Soc., (1949), S163.

4. Hodge, John E., Mills F.D., Fisher, B.E., Cereal
    Science Today, (1972), 17, 34.
5. Rizzi, George P., J. Org. Chem., (1971), 36, 1710.
6. Rizzi, George P., J. Agr. Food Chem., (1974), 22, 279.
7. Chatelus, George, Bull. Soc. Chem. Fr., (1964), 2523.
8. Anet, Edward F.L.J., Chem. Ind. (London), (1962), 262.
9. Burton, H.S., McWeeny, D.J., Biltcliffe, E.O., J. Sci.
    Food Agr., (1963), 14, 911.
10. Sutton, D.A., J. Chem. Soc., (1944), 306.
11. Gensler, Walter J., Rockett, John C., J. Amer, Chem.
    Soc., (1955), 77, 3262.
12. Baddiley, J., Ehrensvard, G., Nilsson, H., J. Biol.
    Chem., (1949), 178, 399.
13. Reeves, Richard L., in "The Chemistry of the Carbonyl
    Group", p. 600, chapter 12, Interscience Publishers,
    N.Y., (1966).
14. Friedman, Mendel, Sigel, Carl W., Biochemistry,
    (1966), 5, 478.

Identification and Flavor Properties of Some 3-Oxazolines and 3-Thiazolines Isolated from Cooked Beef

CYNTHIA J. MUSSINAN, RICHARD A. WILSON, IRA KATZ,
ANNE HRUZA, and MANFRED H. VOCK

International Flavors and Fragrances, Inc., 1515 Highway 36, Union Beach, N. J. 07735

The flavor of meat results from the chemical reactions occurring when a complex mixture of flavor
precursors is heated in an aqueous environment. Attempts to elucidate the specific pathways involved have
shown that while Maillard type reactions are important
for the development of meat flavor, they are only part
of a complex process. Recent work, therefore, has been
increasingly concerned with the isolation and identification of the volatile flavor chemicals from cooked
meat. As these studies continue, more and more heterocycles containing sulfur and/or nitrogen are being
identified. Typical of these are the compounds shown
in Table I which have been reported as constituents of
beef volatiles. Trithioacetaldehyde, trithioacetone,
and thialdine, according to the patent literature (1),
are useful for chicken and beef flavors. According to
Brinkman et al (2), the thialdine "can be considered as
a contributor to beef broth flavor". The 3,5-dimethyl-
1,2,4-trithiolan was reported by Chang et al (3) in
1968 in boiled beef. 1-Methylthioethanethiol was reported in the headspace volatiles of beef broth by
Brinkman et al (2) and was described as having the odor
of fresh onions. The other two mercaptans shown in
Table I were first reported by Wilson et al (4) and
were subsequently patented for use in meat and roasted
meat flavors (5).
    We believe that one of the reasons why more of
these compounds have not yet been reported is that
most of them are present in only trace quantities.
Since these are new compounds, their identification is
even more difficult due to the lack of available reference spectra. The instability evidenced by many of
them may also be a factor.
    It is the purpose of this paper to present the
identification and flavor properties of some new sulfur

TABLE I   CHEMICALS ISOLATED FROM COOKED BEEF

Trithioacetaldehyde

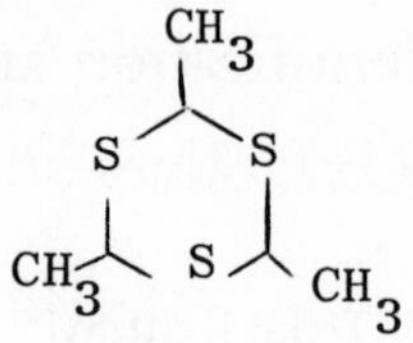

Trithioacetone

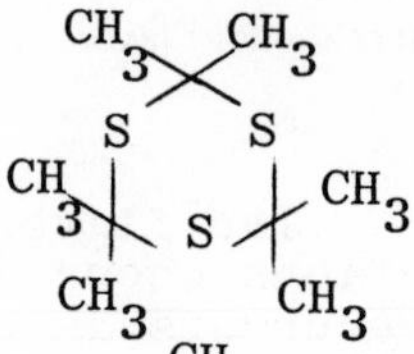

Thialdine

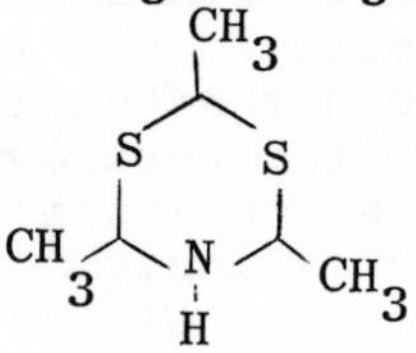

3,5-Dimethyl-1,2,4-Trithiolane

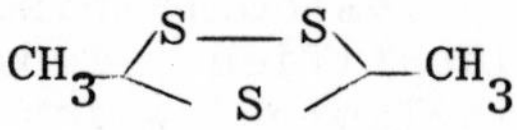

1-Methylthioethanethiol

$$CH_3-S-\underset{\underset{CH_3}{|}}{CH}-SH$$

3-Methyl-2-Butanethiol

$$CH_3-\underset{\underset{SH}{|}}{CH}-\underset{\underset{CH_3}{|}}{CH}-CH_3$$

2-Methyl-1-Butanethiol

$$CH_3-CH_2-\underset{\underset{CH_3}{|}}{CH}-CH_2-SH$$

and nitrogen containing heterocycles, specifically 3-oxazolines and 3-thiazolines, from cooked beef.

The first report of a 3-oxazoline in meat was made by Chang et al (<u>3</u>) in 1968. These workers isolated and identified 2,4,5-trimethyl-3-oxazoline from boiled beef. The volatiles were separated by a combination of flash vaporization and evaporation from a thin heated film and fractionated by preparative gc. The compound was then isolated from one of two fractions that, according to these authors, possessed a "characteristic boiled beef aroma".

CH₃ structures

2,4,5-TRIMETHYL-3-OXAZOLINE   2-ACETYL-2-THIAZOLINE

Tonsbeek et al (<u>6</u>) isolated a compound with the aroma of "freshly baked bread crust" from beef broth. In this work, beef broth was simmered for 2.5 hrs. in a vessel covered with a stainless steel lid. After filtering the slurry, various techniques were used to obtain a fat and fatty acid-free extract from which the desired compound was isolated. This compound was later identified as 2-acetyl-2-thiazoline. The identification of 2,4,5-trimethyl-3-oxazoline and 2-acetyl-2-thiazoline represent the first reported isolation of compounds of this type from foods. The occurrence of the trimethyl oxazoline in beef was later confirmed by Pokorny (<u>7</u>).

The isolation scheme used in this investigation is shown in Figure 1. Forty pounds of lean ground round beef were slurried with water and cooked in a stainless steel vessel at 162.7° for 15 min. A second 40 lb. cook was made at 182°. The lower temperature sample was filtered through cheesecloth. The filtrate was simultaneously atmospherically steam distilled and continuously extracted with distilled diethyl ether in a William's apparatus (<u>8</u>). The 182° sample was atmospherically distilled, and the distillate was saturated with sodium chloride and continuously extracted with distilled diethyl ether. The extract in each case was dried over sodium sulfate and concentrated by careful distillation in a Kuderna-Danish concentrator.

The concentrates were fractionated by preparative gas chromatography and analyzed on a Hitachi Model RMU-6E mass spectrometer coupled with a Hewlett-Packard Model 5750 gas chromatograph. The chromatographic columns used were 500 and 1000' x .03" stainless steel open tubular columns coated with CBW 20M and SF-96.

Further details of the isolation procedures were reported in two previous publications (4), (9).

The compounds were identified by comparing their mass spectra and retention times with those of authentic compounds synthesized in our laboratory.

As an aid in the detection of these heterocycles, the total samples and each trap were analyzed on a Tracor MT 220 gas chromatograph modified to simultaneously detect sulfur and/or nitrogen containing compounds.

The compounds identified and their I_E values, or retention indices relative to a series of ethyl esters of normal alkanoic acids (10), are given in Table II. The 2,4-dimethyl-3-oxazoline was only tentatively identified due to the presence of another component in the spectrum of the unknown. Also found were the trimethyl, 2,4-dimethyl-5-ethyl-and 2,5-dimethyl-4-ethyl-3-oxazolines and the 2,4-dimethyl-and trimethyl-3-thiazolines.

The oxazolines were all found in both preparations while the thiazolines were only identified in the lower temperature sample. It would appear from these data that the thiazolines are less stable to heat than the oxazolines. In 1960, Asinger et al (11) observed that 3-thiazoline can be thermally dehydrogenated in the presence of elemental sulfur to the corresponding thiazole. These thiazoles were, indeed, found in both cooks. Of course, it's also possible that the thiazolines were present in the high temperature sample, but not detected. None of these compounds except 2,4,5-trimethyl-3-oxazoline have ever been reported as constituents of food volatiles.

The 3-thiazolines were prepared according to the method of Asinger et al (12) as shown below for 2,4,5-trimethyl-3-thiazoline. Ammonia was reacted with the appropriate aldehyde and mercapto ketone. After refluxing and steam distilling, the products were ex-

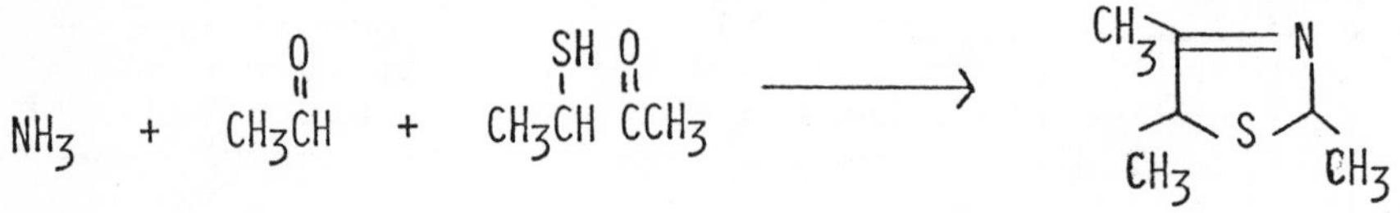

$$NH_3 \quad + \quad CH_3\overset{\overset{\text{O}}{\|}}{C}H \quad + \quad CH_3\overset{\overset{\text{SH}}{|}}{C}H \ \overset{\overset{\text{O}}{\|}}{C}CH_3 \quad \longrightarrow$$

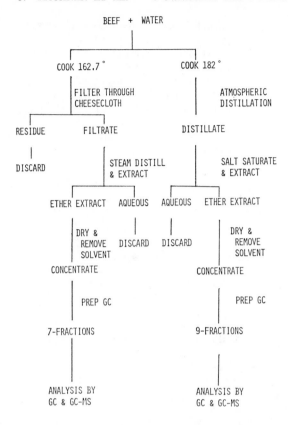

Figure 1. Isolation procedures

TABLE II

3-OXAZOLINES AND THIAZOLINES IN BEEF

COMPOUND	I_E KNOWN [A]	I_E UNKNOWN	COOK
2,4-DIMETHYL-3-OXAZOLINE [B]	3.64 SF-96	3.66	162.7°, 182°
2,4,5-TRIMETHYL-3-OXAZOLINE	4.90	4.81	162.7°, 182°
2,4-DIMETHYL-5-ETHYL-3-OXAZOLINE	5.62	5.64	162.7°, 182°
2,5-DIMETHYL-4-ETHYL-3-OXAZOLINE	5.40	5.41	162.7°, 182°
2,4-DIMETHYL-3-THIAZOLINE	7.50	7.60	162.7°
2,4,5-TRIMETHYL-3-THIAZOLINE	7.35	7.41	162.7°

[A] I_E - VALUES WERE DETERMINED ON A 500 FT. X 0.03 IN. I.D. STAINLESS STEEL OPEN TUBULAR COLUMN COATED WITH CARBOWAX 20M UNLESS OTHERWISE INDICATED.

[B] - THIS COMPOUND WAS ONLY TENTATIVELY IDENTIFIED DUE TO THE PRESENCE OF ANOTHER COMPONENT IN THE SPECTRUM OF THE UNKNOWN.

tracted with ether. The solvent was removed and the
product was purified by preparative gas chromatography.
The oxazolines were synthesized in a similar manner as
described by Jassman and Schulz (13) substituting the
appropriate acyloin for the mercapto ketone. This is
illustrated below for 2,4,5-trimethyl-3-oxazoline.
The reaction mixture was stirred at room temperature,
and the products were extracted with ether. After removing the solvent, the product was purified by frac-
tional atmospheric distillation.

$$NH_3 \quad + \quad \overset{\overset{\text{O}}{\|}}{CH_3CH} \quad + \quad \overset{\overset{\text{OH}}{|}\overset{\text{O}}{\|}}{CH_3CH\,CCH_3} \quad \longrightarrow \quad$$

     Figures 2, 3, and 4 show the mass spectra of the
oxazolines and thiazolines found by us in cooked beef.
The spectra of all compounds except 2,4-dimethyl
oxazoline were obtained on a CEC-103C mass spectrometer
with an ionization voltage of 70eV and source and inlet
temperatures of 150°. The spectrum of 2,4-dimethyl
oxazoline was obtained on a Hitachi RMU-6E mass spectrometer operated under the same conditions.
     We believe that it's possible that these com-
pounds form in beef in a manner similar to that by
which they are synthesized. In the case of 2,4,5-tri-
methyl-3-oxazoline, the starting materials, ammonia,
acetaldehyde, and acetoin have all been reported as
constituents of cooked meat. The same is true for the
corresponding thiazoline. In this case, however, the
alpha mercapto ketones have not been reported in beef
or any other natural product. This is not too surprising
since these compounds are probably highly reactive.
Nevertheless, the mercapto ketones could exist as
transient species formed from available precursors. It
has been suggested and confirmed in our laboratory that
hydrogen sulfide can react with and replace a carbonyl
function. Thus, as shown below, with a dicarbonyl such
as diacetyl which does occur in beef, one could expect
a mixture of products including the thiazoline precursor 2-keto-3-butanethiol. This compound, together with

I $\overset{\overset{\text{O}}{\|}\overset{\text{O}}{\|}}{CH_3C\,CCH_3} \quad + \quad H_2S \quad \longrightarrow \quad \overset{\overset{\text{SH}}{|}\overset{\text{O}}{\|}}{CH_3CH\,CCH_3}$

II $\overset{\overset{\text{SH}}{|}\overset{\text{O}}{\|}}{CH_3CH\,CCH_3} \quad + \quad NH_3 \quad + \quad \overset{\overset{\text{O}}{\|}}{CH_3CH} \quad \longrightarrow \quad$

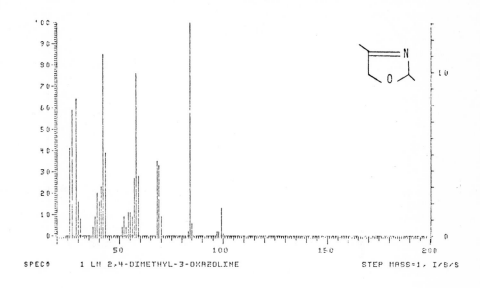

SPEC# 1 LM 2,4-DIMETHYL-3-OXAZOLINE   STEP MASS=1, I/B/S

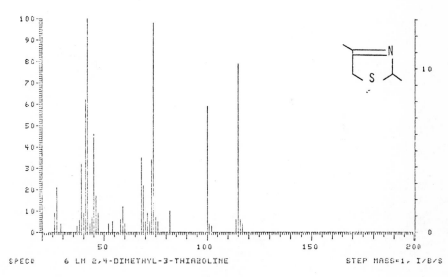

SPEC# 6 LM 2,4-DIMETHYL-3-THIAZOLINE   STEP MASS=1, I/B/S

Figure 2.

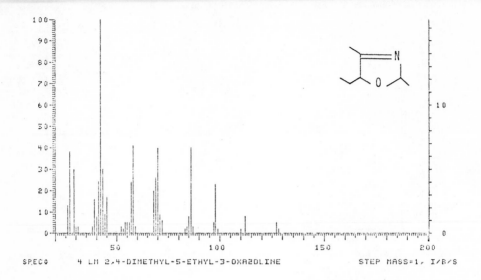

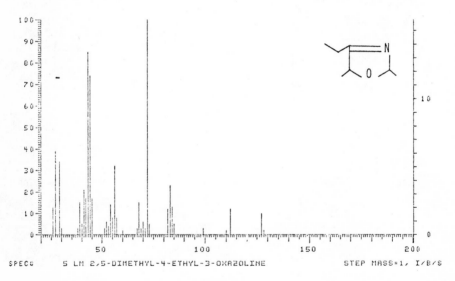

Figure 3.

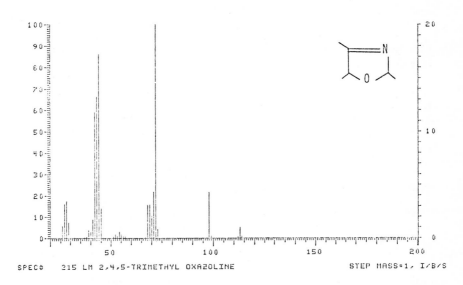

SPEC# 215 LM 2,4,5-TRIMETHYL OXAZOLINE   STEP MASS=1, I/B/S

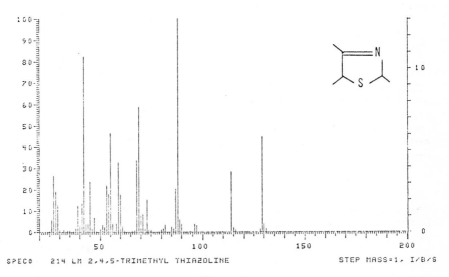

SPEC# 214 LM 2,4,5-TRIMETHYL THIAZOLINE   STEP MASS=1, I/B/S

Figure 4.

the other precursors, ammonia and acetaldehyde, could
react to form 2,4,5-trimethyl-3-thiazoline.

Since ultimately the value of chemicals identified
in natural products depends upon their organoleptic
properties, the remainder of this paper will be concerned with this aspect.

Table III gives the flavor threshold in spring
water, as determined by one of our flavorists, for the
thiazolines, oxazolines, and some structurally related
oxazoles and thiazoles. As expected, the sulfur containing compounds, in general, have lower thresholds,
the lowest being 2,4-dimethyl-5-ethyl-thiazole at
2 ppb.

A panel of expert flavorists assigned the flavor
descriptors shown in Table IV to the thiazolines and
oxazolines and some related compounds. Examination of
this table reveals several interesting correlations.
First, almost all of these closely related compounds
have been described as nutty. That is, all except 2,4-
dimethyl thiazole, 2,4,5-trimethyl-3-oxazoline, and 2-
acetyl-2-thiazoline. Second, most of the sulfur compounds were also described as meaty or roasted while
none of the oxazolines were. This fact together with
their lower thresholds would seem to indicate that
these compounds would be more likely to contribute to
the roasted, meaty character of beef. It should also
be mentioned here that all of the compounds shown in
this table have been found by us in beef volatiles except the 2-acetyl-2-thiazoline. The latter compound,
as previously mentioned, was found by Tonsbeek et al
in beef broth (6). Also apparent from this table is
that most of the oxygenated compounds were found to be
sweet and green in character while none of the sulfur
compounds were so described.

Thus, when we compare the flavor descriptors of
the thiazolines and oxazolines, we find that while 2,4-
dimethyl thiazoline is nutty, roasted, and vegetable,
the corresponding oxazoline was nutty and vegetable,
but not roasted. The 2,4,5-trimethyl-3-thiazoline is
meaty, nutty, and onion-like while its oxazoline ana-
logue is woody, musty, and green. The 2,4-dimethyl-5-
ethyl thiazole is nutty, roasted, and meaty while the
2,4-dimethyl-5-ethyl oxazoline is nutty, sweet, green
and woody.

The thiazoles and thiazolines are obviously much
more closely related to each other than to the oxygenated derivatives. The same, of course, is true for
the oxazolines and the one oxazole studied. Thus,
while many of the sulfur compounds are meaty and
roasted most of the oxygenated compounds are sweet and

TABLE III

THRESHOLD DATA

THIAZOLES	THIAZOLINES	OXAZOLINES	OXAZOLES
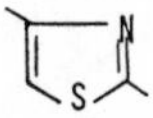 0.1 PPM	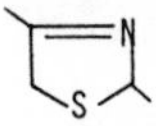 0.02 PPM	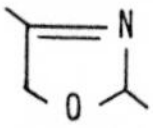 1.0 PPM	
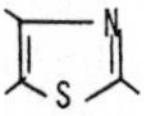 0.05 PPM	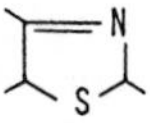 0.5 PPM	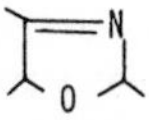 1.0 PPM	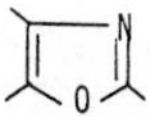 0.005 PPM
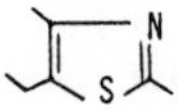 0.002 PPM		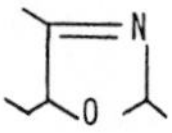 0.5 PPM	
		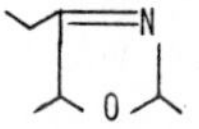 1.0 PPM	

TABLE IV

FLAVOR DESCRIPTORS

THIAZOLES	THIAZOLINES	OXAZOLINES	OXAZOLES
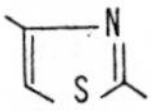 MEATY COCOA-LIKE	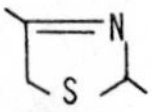 NUTTY ROASTED VEGETABLE	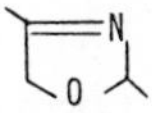 NUTTY VEGETABLE	
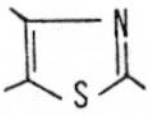 COCOA NUTTY	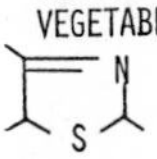 MEATY NUTTY ONION	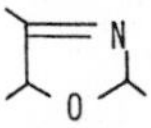 WOODY MUSTY GREEN	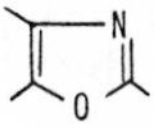 NUTTY SWEET GREEN
NUTTY ROASTED MEATY		NUTTY SWEET GREEN	
		WOODY NUTTY SWEET VEGETABLE	
NUTTY CEREAL POPCORN	BREADY		

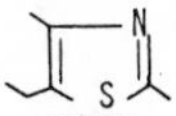

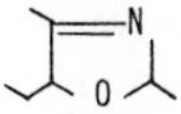

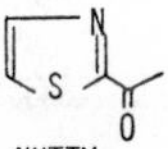

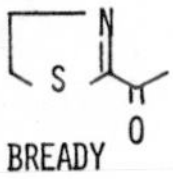

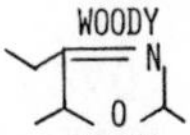

green. The limited number of compounds studied must
be kept in mind in drawing any conclusions from this
data. Nevertheless, it seems to verify the well known
fact that changing one of the hetero atoms of a compound has a much greater effect on its flavor than the
degree of unsaturation it possesses. The 2-acetyl-2-
thiazoline was described by Tonsbeek et al as bready
(6). The corresponding thiazole, while not specifically "bready", was nevertheless described as cereal or
popcorn-like. The presence of the acetyl group apparently makes these compounds more similar to each other
than to other members of their respective classes.
     The data for most of the thiazoles was taken from
a paper by Pittet and Hruza entitled "Comparative Study
of Flavor Properties of Thiazole Derivatives" (14).
These authors compared the flavor properties of 26
thiazole derivatives to those of the corresponding
pyrazines and pyridines.
     The utility of oxazolines and thiazolines as
flavorants is exemplified by the patent literature.
For instance, German patent no. 2120577 (15) concerns
the use of 3-oxazolines to give cocoa, fruit, or
butter-like aromas. German patent no. 2226780 (16)
describes the use of 2-alkyl-3-thiazolines in flavoring
compositions for food, and U.S. patent no. 3778518 (17)
concerns the use of 2-acyl-2-thiazolines to improve the
fried, roast, or baked flavor of foods.
     In summary, cooked beef contains nitrogen and sul-
fur heterocycles among which are thiazoles, thiazolines,
oxazoles, and oxazolines. Based upon their flavor
properties, it is apparent that they contribute to the
over-all flavor properties of cooked beef. As new and
improved analytical techniques and additional reference spectra become available, we are confident that
the identification of many more of these heterocycles
will be forthcoming.

Literature Cited

1. Wilson, R.A., Vock, M.H., Katz, I., Shuster, E.J.,
    British Patent 1364747 (1974).
2. Brinkman, H.W., Copier, H.J., deLeuw, J.J.M.,
    Tjan, S.B., J. Agr. Food Chem. (1972), 20(2), 177.
3. Chang, S.S., Hirai, C., Reddy, B.R., Herz, K.O.,
    Kato, A., Chem. Ind. (London) (1968), 1639.
4. Wilson, R.A., Mussinan, C.J., Katz, I., Sanderson,
    A., J. Agr. Food Chem. (1973), 21(5), 873.
5. Katz, I., Wilson, R.A., Mussinan, C.J., U. S.
    Patent 3713848 (1973).
6. Tonsbeek, C.H.T., Copier, H., Plancken, A.J., J.
    Agr. Food Chem. (1971), 19(5), 1014.
7. Pokorny, J., Prum. Potravin (1970), 21(9), 262.

8. Williams, A.A., Chem. Ind. (London) (1969), 1510.
9. Mussinan, C.J., Wilson, R.A,, Katz, I., J. Agr. Food Chem. (1973), 21(5), 871.
10. van den Dool, H., Kratz, P.D., J. Chromatogr. (1963), 11, 463.
11. Asinger, F., Thiel, M., Dathe, W., Hampel, O., Mittag, E., Plaschil, E., Schroeder, C., Ann. Chem. (1960), 639, 146.
12. Asinger, F., Schafer, W., Herkelmann, H.R., Reintges, B.D., Scharein, G., Wegerhoff, A., Ann. Chem. (1964), 672, 156.
13. Jassmann, E., Schulz, H., Pharmazie (1963), 18(8), 527.
14. Pittet, A.O., Hruza, D.E., J. Agr. Food Chem. (1974), 22(2), 264.
15. Hetzel, D.S., Torres, A., German Patent 2120577 (1971).
16. Dubs, P., Pesaro, M., German Patent 2226780 (1973).
17. Copier, H., Tonsbeek, C.H.T., U.S. Patent 3778518 (1970).

10

Nonvolatile Nitrogen and Sulfur Compounds in Red Meats and Their Relation to Flavor and Taste

AHMED FAHMY MABROUK

Food Science Laboratory, U.S. Army Natick Development Center, Natick, Mass. 01760

Each meat has a characteristic flavor, which can be modified by changing cooking conditions: roasting, pan-frying, boiling, or broiling. The different flavors which arise under these conditions are more easily accounted for. If meat is boiled, the chemical reactions which produce flavors take place at 100°C, but when it is roasted, the temperature may rise to 150°C, or higher. Thus, in addition to the reactions taking place in the hot juices of the meat, the pyroletic reactions occurring at the surface are responsible for the roasted flavor notes.

As it is necessary to heat meat to develop the desired flavors, meat must contain substances which produce these flavors upon cooking. These compounds are called flavor precursors; they may occur naturally in raw meat or may be produced during processing. The non-aqueous flavor precursors (lipids) produce compounds which take part in browning reactions for the development of cooked meat aroma notes. Species aroma differences resulting from this reaction seem to be more in quantity than in quality. While odd-numbered n-fatty acids, and abnormal proportions of branched chain fatty acids are responsible for the characteristic species flavor notes of cooked lamb and mutton (1), 5 -androst-16-ene-3-one is responsible for the taint notes in boar meat (2).

Two approaches are used in meat flavor research. The most common approach is the isolation and identification of volatile flavor components. The other one is to establish the identity of the individual components of flavor precursors. Both approaches have provided indispensable informations. Several hundred compounds had been identified in cooked beef volatiles which had odor or taste reminiscent of cooked beef. These components are not, however, the compounds we would recognize as beef flavors. In this paper, I will be correlating the work done on beef flavor precursors in our laboratory and by other researchers with published data on volatiles collected from cooked meats and from pure components of flavor precursors.

Aqueous meat flavor precursors encompass fifteen classes of organic compounds: 1) glycopeptides, 2) nucleic acids, 3) free nucleotides, 4) peptide-bound nucleotides, 5) nucleotide sugar, 6) nucleotide sugaramine, 7) nucleotide acetylsugaramine, 8) nucleosides, 9) peptides, 10) amino acids, 11) free sugars, 12) sugar phosphate, 13) sugaramine, 14) amines, and 15) organic acids. Compounds belonging to these fifteen sets contribute flavor notes to the overall impression of cooked meat.

All raw red meats have weak blood-like flavor, on which was superimposed flavor notes distinctive of species, food, and environment of the animal (3). The flavors developed on cooking were similar for the various meats but again modified by species, food, and environment of the animal. These characteristics tended to be lost with prolonged cooking. Such differences seemed to be more in quantity than in quality. All red meats were found to have some of the flavor characteristics of fish, and birds; and the meat of birds and fish had red meat characteristics.

The dialyzable water-soluble fraction of ox muscle was reported to contain primarily amino acids and reducing sugars (4). When a mixture of these identified amino acids was heated with glucose, a meaty aroma and flavor was produced (5). Omission of glucose resulted neither in browning color nor in flavor development, thus browning-type reaction may be responsible for lean meat flavor production. Some of beef flavor precursors have been reported to be a relatively simple mixture of glucose, inosinic acid, and a glycoprotein (6, 7). Glutamic acid, serine, glycine, alanine, β-alanine, isoleucine, leucine, and proline were the amino acid components of the glycoprotein. When these amino acids were used in conjunction with glucose, inosine, and inorganic phosphate, meaty odors and flavors were produced upon heating. The volatiles from lean meats such as beef, pork, and lamb were found to contribute an identical meaty flavor, and that species' flavor differences can be traced to the fat content (8). Thirty one amino compounds were reported in water extracts of beef, pork, and lamb (9, 10). Methionine was present in the meaty aroma fractions obtained by dialysis and gel permeation chromatography of aqueous beef extracts (11). A study to evaluate the individual contribution of the nitrogenous components of beef flavor precursors indicated that tyrosine, phenylalanine, taurine, and glutamic acid may be removed from the flavor fractions without seriously affecting the aroma (12). Furthermore, creatine, creatinine, purine, and purine derivatives are not involved in the development of meaty aroma. Beef flavor precursors were found to be distributed among seven major fractions obtained by dialysis (or ultrafiltration) and gel permeation chromatography of aqueous extracts (13, 14). The methods employed were sufficiently precise to permit distinction between different muscles. Methionine and cysteic acid were found to be the most important amino acids contributing to flavor. Evidence suggests that precursors characteristics reside in more than one molecular structure and are correlated with sulfur containing amino acids present. The complexity of meat flavor precursors will be clear from the following discussion on the unpublished data from our laboratory on the glycopeptides fraction (15).

Glycopeptides

An aqueous solution of (ca. 20g.) freeze dried raw beef diffusate prepared according to Mabrouk et al.(13) was fractionated by preparative gel permeation chromatography (PGPC) on three columns (4.5X110 cm) of Sephadex. The void volume from Sephadex G-10 column was passed through Sephadex G-15 column. The excluded fraction from the second column was fractionated on G-25F column. All resulting fractions were freeze dried and their odor upon heating were evaluated. The second fraction from G-10 column gave an intense meaty odor when heated. The total nitrogen and carbohydrate contents of this fraction are 7.96-0.23 and 9½% (calculated as glucose), respectively. Quantitative data on its amino compounds content (mg./100 g.) are: glucosamine (1.11), hydroxylysine (0.19), lysine (0.44), histidine (1.10), threonine (2.54), serine (3.96), glutamic acid (21.30), proline (1.92), glycine (5.00), arginine (1.00), phosphoserine (trace), taurine (9.17), methionine sulfoxide (2.16), aspartic acid (3.94), α-alanine (16.00), valine (5.64), methionine (3.55), isoleucine (4.81), leucine (9.74), phenylalanine (2.01), β-alanine (0.76), and two unknowns. Upon further fractionation of this fraction on an analytical column (1.1 X 110 cm) of Sephadex G-25F , it was resolved to four peaks (Figure 1). While the second and third fractions had intense meaty odor, the first and fourth did not. Attempts to improve the fractionation by using Sephadex G-10 (1.1 X 110 cm) were unsuccessful as none

of the resulting five fractions developed the pleasant
meaty odor upon heating. While thin layer chromatography indicated the presence of at least seven compounds
in PGPC beef flavor precursor, thin layer electrophore-
sis showed only three resolved spots upon reacting with
ninhydrin.

To provide sufficient material for fractionation
and sensory evaluation, a preparation of beef ultra-
filtrate (14) followed by preparative gel permeation
chromatography was achieved (Figure 2).

The meaty fraction obtained by PGPC was fractionated by preparative thin layer chromatography(PTLC),us-
ing MN 300 cellulose plates, 20X20 cm., precoated (ca.
500-micron thickness). About 200-μl of 0.075 g./ml.so-
lution of this fraction was applied to each plate. The
plates were developed in n-propanol-water, 7:3(v./v.),
dried,visualized and scraped (Figure 3). The resulting
PTLC cellulose fractions were extracted twice with 50%
ethanol, the alcohol extracts were concentrated under
vacuum at room temperature, freeze dried and stored in
a dessicator under vacuum above phosphorus pentoxide at
-20°C. The nitrogen content and odor evaluation of each
PTLC fraction are listed in Table I.

The resulting seven PTLC fractions were hydrolyzed
by ion exchange resin catalysis following the procedure
of Paulson et al (16). The amino acids present in these
fractions are recorded in Table II. The major amino
acids in each fraction are listed in Table III.

Table I indicates that PTLC fraction # 2 exhibited
strong meaty aroma intensity. When this fraction was
subjected to preparative TLC, eight subfractions resulted (Figure 4). About 200-μl of 0.075 g./ml. solution of this fraction was applied to 108 preparative
TLC plates MN 300 cellulose (ca. 500-micron thickness).
Thus, a total of 1.62 g. of TLC fraction # 2 was submitted to a second step of fractionation by PTLC.
Eight PTLC sufractions were collected. Table IV lists
the weight of each subfraction recovered. The excess
recovery of 0.82687 g., appears to be due to the presence of materials in the plates which were soluble in
the 50% ethanol used to extract each subfraction. PTLC
plates used in fractionation proved to contain materials that were extracted with ethanol during recovery of
the subfractions. The coatings of ten plates were
scraped and extracted with 50% ethanol. About 60 mg.
"contaminants" were recovered, i.e., about 6 mg./plate.
The chromatograms obtained with analytical gel permeation chromatography using Sephadex G-10 (0.7X28 cm.) of
the contaminants (Figure 5) and PTLC subfraction # 3
(Figure 6), indicate the feasability of separating the

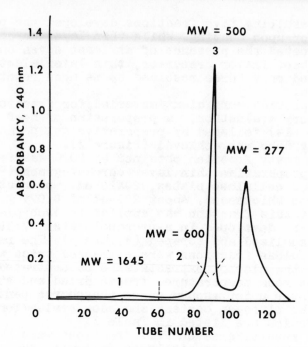

Figure 1.   Gel permeation chromatography of GPC-BFP,
110 × 1.1 cm column, Sephadex G25F

		% OF FRESH BEEF
1— TRIM, GRIND, BLEND WITH WATER (1 LITER /KG.)		
2— FREEZE DRY, LYOPHILIZED BEEF		28.20
3— EXTRACT WITH PETROLEUM ETHER, LIPIDS		3.50
4— BLEND FREE—LIPIDS FREEZE DRIED BEEF WITH WATER,(10 ML. / G.), STIR SLURRY FOR THREE HOURS AT 5°C, CENTRIFUGE ,FILTER		
5— REPEAT STEP 4, FREEZE DRY FILTRATES		6.70
6— DIALYSIS		
a—ULTRAFILTRATION		3.68
b—CONVENTIONAL		2.98
7— PREPARATIVE GEL PERMEATION CHROMTOGRAPHY		1.66
(BROILED BEEF AROMA WHEN HEATED)		

Figure 2.   A schematic outline of the method used to prepare aqueous beef flavor
precursors

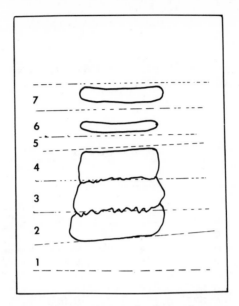

Figure 3.   Preparative thin layer chromatogram of PGPC-BFP, MN 300 cellulose plate, 500 μ thickness, developing solvent: propanol–water, 7:3 (v/v), silver nitrate visualization

Table I.   Comparison of PTLC Fractions of PGPC of Beef Flavor Precursors

Fraction No.	Nitrogen Content %	Yield % of PGPC	Odor Description
1	4.42	)	BFP, Very Sweet.
		)29	
2	9.87	)	Strong BFP, Very Sweet.
3	13.13	)	Weak BFP,Oily,Not Sweet.
		)45	
4	7.03	)	Weak BFP, Sweet.
5	4.28	10	Heated Protein,Fishy,Musty, Not Sweet.
6	3.52		Wet Cardboard.
7	2.88		Musty, Hay.
PGPC-BFP	7.92		
Ultrafiltrate	9.20		

Table II.   Amino Acids Present in PTLC Fractions of PGPC-Beef
            Flavor Precursors.

Amino Acids	µM/mg. of Fraction No.						
	1	2	3	4	5	6	7
Histidine	0.]26	0.023	0.003	-----	0.001	-----	-----
Glutamic	0.098	0.299	0.025	0.002	0.003	0.003	0.0002
B-alanine	0.110	0.026	0.008	-----	0.003	-----	-----
Aspartic	0.052	0.023	0.004	0.001	0.003	0.001	0.0004
Glycine	0.059	0.281	0.020	0.003	0.003	0.012	0.0001
Alanine	0.040	0.078	0.503	0.207	0.008	0.006	0.0001
Arginine	0.017	0.014	-----	-----	-----	-----	-----
Serine	0.022	0.171	0.023	0.003	0.008	0.006	0.0004
Tyrosine	0.009	0.001	-----	-----	-----	-----	-----
Leucine	0.010	0.003	0.003	0.004	0.051	0.427	-----
1-methylhisti- dine	0.008	0.004	-----	-----	-----	-----	-----
Tryptophan	0.005	0.023	0.041	0.017	0.001	-----	-----
Threonine	0.008	0.020	0.066	0.011	0.004	0.002	0.0001
Lysine	0.006	0.003	-----	-----	0.001	-----	-----
Valine	0.006	0.006	0.006	0.014	0.413	0.009	-----
Isoleucine	0.004	0.002	0.002	0.002	0.047	0.223	-----
Ammonia	0.024	0.212	0.091	0.019	0.020	0.021	0.005
Ornithine	0.001	0.002	-----	0.001	-----	-----	0.0001
Phenylalanine	0.001	0.001	0.001	-----	0.006	0.003	-----
α-amino-n-buty- ric acid	0.001	0.003	-----	-----	-----	-----	-----
Proline	-----	-----	0.010	0.048	0.005	0.002	-----
Methionine) Methionine) Sulfoxide)	-----	-----	0.002	0. 001	0.041	-----	-----
Cysteic Acid	-----	-----	-----	-----	0.001	-----	-----

Table III.   The Major Amino Acids in PTLC Fractions

Amino Acids	μM/mg. of Fraction No.					
	1	2	3	4	5	6
Histidine	0.126					
Glutamic	0.098	0.299				
B-alanine	0.110					
Aspartic	0.052					
Glycine	0.059	0.281				
α-alanine	0.040	0.078	0.503	0.207		
Arginine	0.017					
Serine		0.171				
Ammonia			0.091			
Tryptophan			0.041			
Proline				0.048		
Valine					0.413	
Leucine					0.051	0.427
Isoleucine					0.047	0.223
Methionine + Methionine Sulfoxide					0.041	
Threonine			0.066			

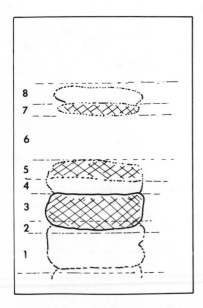

Figure 4.   Preparative thin layer chromatogram of PTLC, fraction #2, MN 300 cellulose plate, 500 μ thickness, developing solvent: propanol–water, 7:3 (v/v), ninhydrin visualization

Table IV.   Weight of PTLC Subfractions of PTLC Fraction 2

Subfraction No.	Weight g.	% of Recovered Material
1	0.31091	12.7
2	0.47833	19.6
3	0.33921	13.9
4	0.18154	7.4
5	0.24900	10.2
6	0.41012	16.8
7	0.25748	10.5
8	0.22028	9.0
Total	2.44687	

Weight of PTLC Fraction #2 used = 1.62 g.

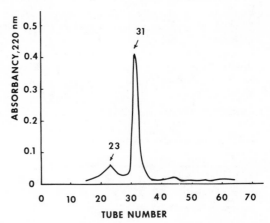

Figure 5.   Gel permeation chromatography of "contaminants," 28 × 0.7 cm column, Sephadex G-10

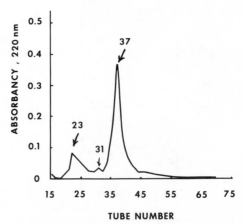

Figure 6.   Gel permeation chromatography of PTLC subfraction #3, 28 × 0.7 cm column, Sephadex G-10

subfractions from the contaminants. The homogeneity of
each subfraction needs to be evaluated before subject-
ing to sensory evaluation and analysis.

Amino Acids

Of 27 amino compounds in aqueous beef extract, 23
were identified and quantified (mg./100 g. fresh weight
meat): phosphoserine,1.84; taurine,0.09; aspartic,
0.95; threonine,3.17; serine,3.60; glutamic,11.60; proline,0.94; glycine,3.10; ∝-alanine,18.17; cystine,0.60;
valine,6.90; methionine and methionine sulfoxide (as
methionine) ,6.48; isoleucine,5.07; leucine,9.73; tyrosine,5.44; phenylalanine,6.04;β-alanine,0.96; glucosamine,3.04; hydroxylysine,0.59; lysine,2.40; anserine,
29.43; and arginine,1.53 (17). The free amino nitrogen
compounds of beef,pork, and lamb are remarkably similar
but differ quantitaively (9, 10). Beef contained the
largest amount of the free amino compounds (161.53mg./
100g. fresh weight), pork the least amount (109.86mg./
100g.), and lamb components totalled 130.01mg./100g.
Table V shows the relative amounts of amino acids in
lyophilized diffusate of beef, lamb, and pork. The ratios of the different free amino compounds in red meats
vary according to the species (Table VI).

Each class of amino compounds (neutral, aromatic,
heterocyclic, hydroxy, sulfur-containing, acidic, and
peptides) most probably make specific contributions to
the flavor of red meats. Lamb content of acidic amino
acids (glutamic, and aspartic) is approximately double
that in either beef or pork. Lamb is also rich in
sulfur-containing amino acids; its content is twice
that of pork and three times the value in beef. The
percentage of basic amino compounds in lamb is three
times that in pork and twice the value of beef. During
processing, each class of amino compounds produces specific volatiles due to non-enzymic browning reactions,
pyrolysis, etc. The data in Tables V and VI, as well as
the variation in the pH values of red meats—beef (5.5-
6.0), pork (6.3), lamb (7.0)— is suggestive that the
volatiles of processed lamb, pork, and beef will differ
either qualitatively or quantitatively or in both. It
is possible that a relationship could be found between
the relative amounts of certain amino compounds and
meat flavors as represented by species. The distinctive
taste of beef and insipid taste of veal is rationalized
by the increase in the content of specific amino acids
(valine, methionine, leucine, isoleucine, lysine, threonine, and tryptophan) with age (18). Also, the values
of the individual essential amino acids and hydroxypro-

Table V.   Relative Amounts of Amino Compounds in Lyophilized Diffusate from Beef, Lamb and Pork.

Amino Compound	Beef	Lamb	Pork
Phosphoserine	36	36	36 (1)
Glycerophosphoethanolamine	2	1	2
Phosphoethanolamine	66	115	45
Taurine	905	3071	1258
Urea	1	tr(2)	---
Aspartic	82	178	137
Threonine	111	367	48
Serine+Asparagine(as serine)	753	555	295
Glutamic	463	711	195
Proline	---	357	64
Glycine	240	503	275
Alanine	1128	1074	419
Cystine	437	756	211
Valine	299	51	30
Methionine	201	277	69
Isoleucine	204	326	103
Leucine	381	658	168
Tyrosine	185	232	56
Phenylalanine	136	216	51
Ornithine	---	111	tr
NH_3 + Lysine	619	1230	427
Histidine	410	1135	255
Anserine+Carnosine (as anserine)	9014	2989	6794
1-methylhistidine	480	242	49
α-amino-n-butyric acid	---	tr	---

(1) Arbitrary Units.
(tr) To Represent Trace Amounts Only.

line differ with respect to species of the animals. Furthermore, the most expensive cuts, such as fillet and round cuts, contain the greatest amount of these amino acids.

Beef, pork, lamb, and mutton from animals in poor condition produced more labile sulfur in the form of hydrogen sulfide than did the meats from animals in good condition (19). The amount of hydrogen sulfide produced was directly related to the pH of the meat, irrespective of whether the pH was varied artificially by the use of acid, or alkali or whether the ultimate pH was increased biologically by starvation.

Free creatine and creatine phosphate are by far the most abundant extractive compounds of muscles as they constitute about 0.5% of fresh muscle. Most of the creatine is present as phosphate ester in resting muscles. Creatine phosphate is an important phosphorylating agent and is concerned in the chemical processes of muscle contraction. The means for creatine content in ox, sheep, and pig muscles are 2.13, 2.06, and 2.11 % of crude protein, respectively (20). While creatine has little taste, creatinine has a markedly bitter taste. In pure solutions, the bitterness is only noticeable at double the concentration present in meat, when conversion to creatinine is complete (21).

Peptides

Red meat peptides may be as complex as those obtained by partial hydrolysis of a protein, i.e., there may be hundreds of components. The separation of pure individual compounds from such a mixture is obviously difficult, but it could be achieved. With the exception of reports (4,9,10) on the presence of carnosine, anserine, and glutathione, no published work exists in literature on the presence of other peptides in red meats. As the total content of the amino acids identified in aqueous meat extracts account for a small portion of the total nitrogen content, the balance of the nitrogenous compounds should account for peptides, glycopeptides, nucleotides, nucleotide derivatives, and amines. The increase in the number of the components obtained by fractionation of meat flavor precursors by PGPC, PTLC, and analytical GPC , indicates the presence of peptides and/or glycopeptides depending upon the presence of carbohydrate moieties (Figures 3 and 4) and (Tables II, III and IV).

Nucleotides and Nucleotide Derivatives

Nucleotides present in red meats vary according not only to species, breed, age of the animal, and feed but also to the freshness of the meat. 5'-Inosine monophosphate (IMP) and 5'-guanosine monophosphate(GMP) have flavor enhancing effects, and these effects are considered to be essentially the same. The flavor effect of GMP is said to be broader than that of IMP and generally produces a more harmonizing effect. GMP is about 4 times stonger than IMP in aqueous solutions. Meats contain a very small quantity of GMP in addition to IMP, but the level is so small that it cannot be considered to have any effect on taste(22). Eight fractions of peptide-bound nucleotides were reported in beef, one fraction exhibited meaty aroma when heated (23). Red meats contain nucleotide sugars which are closely related to the metabolism and biosynthesis of sugars. Nucleotide choline is involved in lipid metabolism. 5'-cytidine monophosphate, 5'-adenosine monophosphate, and 5'uridine monophosphate are also present in meat in small concentrations. Nucleotide content in red meats is tabulated in Table VII.

Effect Of Heating On Meat Flavor Precursors

During heat processing, the types of reactions which can occur to the non-aqueous flavor precursors are: autoxidation, hydrolysis, dehydration and decarboxylation of fats giving rise to aldehydes, fatty acids, lactones, ketones, hydrocarbons, alcohols,..etc. The aqueous flavor precursors may be subjected to glycoside splitting, oxidation, pyrolysis (thermal decomposition) to yield volatile and non-volatile compounds that influence flavor. Products of amino acids pyrolysis are very complex. The most important flavor producing chemical reaction is the non-enzymic browning reaction of the amino acids. This reaction involves oxidative deamination of an amino acid molecule with the formation of an aldehyde with one carbon atom less than the original acid.

a- Nucleotides. In a study on the browning reaction of 5'-ribonucleotides with D-glucose,Fujimaki et al. (26) concluded that phosphate plays an important role in the browning reaction of aldoses with nucleotides, and that the phosphate ester at the primary alcohol of ribose residue, as well as, the inorganic orthophosphate which is liberated due to the hydrolysis of the ester causes the development of the reaction. At

Table VI. Classification of Free Amino Compounds in Red Meats.

Amino Compound	Percent		
	Beef	Lamb	Pork
Neutral Amino Acids	14.4	18.1	9.4
Hydroxy Amino Acids	5.5	6.2	3.4
Acidic Amino Acids	3.4	5.9	2.9
Sulfur-Containing Amino Acids	9.6	27.0	13.9
Basic Amino Acids	9.8	20.2	7.1
Aromatic Amino Acids	2.0	2.9	1.0
Dipeptides "Anserine + Carnosine"	55.8	19.7	61.1

Table VII. Nucleotide Content in Red Meats

Meat	Nucleotide Content, mg/100g.					Reference
	CMP	UMP	IMP	GMP	AMP	
Beef	12.0	13.0	150.0	8.0	17.0	24
Beef	1.0	1.6	107.0	2.1	6.6	22
Pork	1.9	1.6	123.0	2.5	7.6	22
Mutton	1.9	0.6	83.5	5.1	6.8	25

higher temperature (120°C), glucose accelerate the degradation of IMP, which may be attributed to interaction of the nucleotide with some reactive compounds
such as osones and others which are possibly formed
through 1,2- and/or 2,3-enolization of aldoses and further degradation, condensation, and polymerization. Although IMP is more stable than GMP, IMP-glucose solution produced more 3-deoxy-D-glucosone and developed
more intense brown color than GMP-glucose solution did,
these differences may be attibuted to the amino group
on the purine ring in GMP. In a study on the effect of
heating on nucleotides in beef, pork, and lamb, Macy
et al (27) reported the following conclusions. In the
case of beef, CMP was not greatly influenced by heating to 49°C and 77°C internal temperature. AMP content
appreciably increased upon heating to both temperatures, this increase might be due to the hydrolysis of
adenosine diphosphate(ADP) and adenosine triphosphate
(ATP). UNP and IMP contents decreased by heating to
49°C and 77°C. The influence of roasting to 49°C and
71°C internal temperatures, on pork nucleotides was
similar to that on beef. All nucleotides decreased
during roasting except AMP which increased in quantity
when heated to 71°C. When lamb was heated to 60°C internal temperature, GMP and UMP were destroyed after 5
and 15 minutes respectively. CMP doubled its original
concentration and then degraded gradually. About 85%
of IMP content was lost after heating for 30 min.(Figure 7).

   b- Amino Compounds and Guanidines. Cooking caused significant increases in creatine and decrease in
amino acids, creatine, non-amino nitrogen and total
carbohydrates (28, 29). In the case of beef,creatine
content loss was dependent on the heating temperature.
The loss amounted to 10 and 20% of the original concentration, when heated up to 49°C and 77°C, respectively. During cooking to 77°C, all free amino acids
of beef roasts increased, except threonine, serine,
glutamic, histidine, and arginine, which decreased. The
overall effect of cooking was an increase of 38.3% in
the total amino acids. Total extractable amino nitrogen
increased approximately 18% in the same sample during
cooking. These increases were probably due to protein
hydrolysis and possibly involved cathepsins or other
proteolytic enzymes in the tissue, since it has been
shown that most free amino acids decreased during heating when isolated from the proteins by dialysis(Table
VIII).Taurine, anserine, and carnosine increased to the
greatest extent during cooking. These were the major

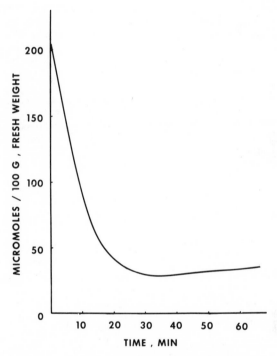

Figure 7.   *Effect of heating lamb to 60°C internal temperature on inosine 5'-monophosphate*

Table VIII. Percent Change in Free Amino Compounds of
Lyophilized Red Meats Diffusates

Amino Compounds	Percent Change		
	Beef	Lamb	Pork
Phosphoserine	- 19.4	- 26.7	- 34.3
Glycerophosphoethanolamine	- 50.0	-100.0	- 50.0
Phosphoethanolamine	+ 24.2	+ 65.2	+320.0
Taurine	-55.5	- 37.3	- 36.9
Urea	-100.0	------	------
Aspartic	- 61.0	- 10.5	- 67.2
Threonine	- 33.3	- 56.7	- 6.3
Serine+Asparagine	- 82.1	- 52.3	- 82.0
Glutamic Acid	- 52.1	- 53.1	- 39.0
Proline	------	- 72.8	- 62.5
Glycine	- 45.0	- 47.6	- 42.5
Alanine	- 44.9	- 28.9	- 33.2
Cystine	-100.0	- 18.9	- 57.8
Valine	- 50.8	- 59.1	- 33.3
Methionine	- 62.7	- 35.4	+ 44.9
Isoleucine	- 57.4	- 33.0	- 19.4
Leucine	- 38.6	- 45.6	- 40.5
Tyrosine	- 56.8	- 15.7	- 33.9
Phenylalanine	- 28.7	- 39.5	- 23.5
Ornithine	------	- 31.6	------
NH_3 + Lysine	- 33.6	- 46.8	- 4.9
Histidine	+ 1.7	- 26.2	+ 22.0
Carnosine + Anserine	- 57.7	- 36.3	- 14.1
1-Methylhistidine	- 89.8	- 47.8	-100.0
αAmino-n-butyric acid	------	------	------
Total	-55.77	- 37.3	- 19.8

(-) decrease in value
(+) increase in value
From reference (9)

amino compounds present in both raw and cooked beef.
The increase in free amino compounds content during
cooking is important for meat flavor development due to
their participation in the browning reaction. The recommended usage of carnosine in soup preparation and
creatine in meat extracts are due to their contribution
the mouthfeel. Anserine is desired as its flavor lin-
gers in the mouth and was therefore savored more
strongly than would be expected simply from a consideration of taste   intensity as measured by threshold
dilution technique (30). Arginine,lysine and histidine
exert the same effect as the flavor sensation clung
tenaciously to the mouth. The odor and taste of cooked
beef were improved by contracting raw beef with basic
amino acids. Smearing beef with arginine before broiling or roasting, resulted in a superior organoleptic
evaluation (31).

Seventeen amino acids and ten peptides were subjected to pyrolysis (Barber Coleman unit Model 4180)
and the products were identified by mass spectrometry,
Table IX (32). Benzene, toluene, ethyl benzene, and
styrene were identified in the products of the thermal
degradadation of phenylalanine (33, 34). These reports
were results of work done at very high temperatures
above 700°C., which is far higher than the temperature
of the oven used in cooking.Usually, the oven used for
roasting is set between 176° and 190° C. During broiling, the temperature at the surface of the meat might
reach a degree as high as 280-300°C. depending upon its
distance from the source of heat. Thus, my discussion
will be limited to reports in the literature where the
temperature did not exceed 300°C.

c- Pyrolysis of Sulfur Containing Amino Acids.
Pyrolysis of cysteine and cystine resulted in 7-8 volatile compounds including 2-methylthiazolodine which is
considered to be a a product of the reaction of acetaldehyde and mercaptethylamine(35). Eleven compounds were
identified in the products of methionine pyrolysis .
Beside these volatile compounds, alanine,cystine and
isoleucine(non-volatiles);and alanine,isoleucine, and
methionine were detected in the pyrolyzed products of
cysteine and cystine, respectively, but no amino acids
were found among methionine products. The mixture of
the seven identified volatile compounds produced from
cystine developed a pop-corn like aroma with a roasted
sesame-like one. Methyl mercaptan seemed to be the main
contributor to a pickled radish odor produced from the
pyrolysis of methionine. In 1973, Kato et al.(36) identified fifteen new volatile compounds in the pyrolysis

Table IX. Pyrolysis Products of Amino Acids and Dipeptides

Compound	Product
Glycine	Acetone
Alanine	Acetaldehyde
B-alanine	Acetic Acid
Valine	2-Methyl propanal
Norvaline	n-Butanal
Leucine	3-Methyl butanal
Isoleucine	2-Methyl butanal
Serine	Pyrazine
Threonine	2-Ethylethyleneimine
Taurine	Thiophene
Methionine	Methyl propyl sulfide
Cystine	Methyl thiophene
Phenylalanine	Benzene
Tyrosine	Toluene
Tryptophan	Ammonia, Carbon dioxide
Proline	Pyrrole
Hydroxyproline	N-Methyl pyrrole
Glycyl-glycine	Acetone
Glycyl-valine	Acetone, 2-Methyl propanal
Glycyl-proline	Acetone, Pyrrole
Glycyl-methionine	Acetone, Methyl propyl sulfide
Glycyl-serine	Acetone, Pyrazine
Glycyl-tryptophan	Acetone, Ammonia
Glycyl-alanine	Acetone, 2-Methyl pyrrole
Alanyl-glycine	Acetone, Acetaldehyde, ammonia
Glycyl-leucine	Acetone, Cyclopentane
Leucyl-glycine	Acetone, Acetic Acid

From reference (32)

products of cysteine and five more from cystine. The popcorn-like aroma most probably is made up from the aroma of many compounds such as hydrogen sulfide,thiazoles, pyridines and 2-methylthiazolidine. Thiazoles and related compounds give a pyridine- or picoline-like odor. α-Picoline gives a mild pyridine-like aroma and resembles popcorn-like one produced from pyrolyzed sulfur containing amino acids (35). Thiophenes were detected in the volatiles of pyrolyzed cysteine,but were not detected in those of pyrolyzed cystine,they generally give slightly unpleasant and characteristic odors. The differences in odor between cysteine and cystine may be partly attributed to the presence of thiophenes in the volatiles. Table X lists the compounds identified in pyrolysis products of sulfur-containing amino acids. Thiazoles and thiazolines may be produced by the oxidation of 2-methyl thiazolidine; and α-picoline and 2-ethyl-5-methyl pyridine may produced by the reaction of acetaldehyde and ammonia. 2-Methyl thiazolidine is thought to be produced from cysteamine and acetaldehyde (37). It is quite obvious that simple compounds such as ammonia, acetaldehyde, cysteamine appear to be important precursors of many volatile compounds.

   d- Pyrolysis of Aromatic Amino Acids. Amine-like, phenol-like and indole-like odors developed from pyrolyzed phenylalanine, tyrosine and tryptophan,respectively (38). Twelve compounds, many of which have aromatic rings, were identified in the volatiles from thermal degradation of phenylalanine(Table XI). Tyrosine and tryptophan produced some phenols and indoles, respectively, along with several other compounds.

   e- Pyrolysis of Hydroxy Amino Compounds. Ten volatile compounds including several pyrazines were identified in the pyrolysis products of L-serine(39). Pyrazines were also identified in the pyrolysis products of L-threonine, but not from alanine and are con sidered to be characteristic pyrolysis products of β-hydroxy amino compounds. Also, diketopiperazines. amines and carbonyl compounds were identified,Table XII. Pyrazines are heterocyclic nitrogen compounds which contribute significantly to the desirable unique flavor and odor associated with roasting or toasting of foods. The odor of pyrazines has been described as characteristically earthy, nutty and roasted. Maga and Sizer(40) reviewed pyrazines in foods. Pyrazines could not be found when glycine, alanine, phenylalanine, β-alanine, leucine, isoleucine,valine, methionine, cystine, hydroxylysine, tyrosine, histidine, proline, hydroxyproline,

Table X.   Pyrolysis Products of Sulfur-Containing Amino Acids

Compounds	Cystine	Cysteine	Methionine
Ethylamine	35	35	35
Propylamine	-	-	35
Allylamine	-	-	35
Crotylamine	-	-	35
α-Alanine	35	35	-
Isoleucine	35	-	35
2-Amino-4-methylthiobutyric Acid	35	-	-
2-Methylthiazolidine	35,36	35,36	-
Mercaptehylamine	35	35	-
Hydrogen Sulfide	35	35	-
Sulfur	35	-	-
3-Methylthiopropylamine	-	-	35
Methional	-	-	-
Acetaldehyde	-	-	35
Propionaldehyde	-	-	35
Isobutyladehyde	-	-	35
Acetone	-	-	35
Ammonia	35	35	-
Ammonium carbonate	35	35	-
2-Methylthiazoline	36	36	
α-Picoline	36	36	
2-Ethyl-5-methylpyridine	36	36	
2-Ethylthiazole	-	36	
Thiophene	-	36	
2-Methylthiophene	-	36	
3-Methyltetrahydrothiophene	-	36	
2,5-Dimethylthiophene	-	36	
2,3-Dimethylthiophene	-	36	
2(or 3)-Ethylthiophene	-	36	
2,3-Dihydro-4(or 5) ethylthiophene	-	36	
2-Methyl-3(or 4)-ethylthiophene	-	36	
2,3,5-Trimethylthiophene	-	36	
3-Methyl-n-propylthiophene	-	36	
2,4-Dimethyl-5-ethylthiophene	-	36	
2-Methylthiazole	36		
2-Methyl-5-ethyl thiazole	36		

From reference (35) and (36)

Table XI. Volatile Compounds Produced From The Pyrolysis of
Aromatic Amino Acids

Volatile Compounds	Phenylalanine	Tyrosine	Tryptophan
Benzylamine	+	-----	-----
Bibenzyl	+	-----	-----
p-Cresol	-----	+	-----
m-Cresol	-----	+	-----
Ethyl Benzene	+	-----	-----
β-Ethylindole	-----	-----	+
Indole	+	-----	+
Phenol	-----	+	-----
β-Phenylethylamine	+	+	-----
β-Phenylpropionitrile	+	-----	-----
Skatole	-----	-----	+
Stilbene	+	-----	-----
Toluene	+	-----	-----
Vinylbenzene	+	-----	-----
Phenylacetaldehyde	+	-----	-----
Acetaldehyde	+	-----	-----
Benzaldehyde	+	-----	-----
Ammonia	-----	+	-----
Methylamine	-----	+	+
Aniline	-----	+	-----
Tyramine	-----	+	+
Ethylamine	-----	-----	+

From reference (38)

Table XII. Volatile Compounds Produced From Hydroxy Amino Compounds

Compound	Serine	Threonine	Ethanol-amine	Glucos-amine	4-Amino-3-hydroxy-butyric	Alanyl-serine
Pyrazine	39,41		41	41		41
2-Methylpyrazine	39,41	41	41	41		41
2,3-Dimethylpyrazine	41			41		
2,5-Dimethylpyrazine		39				
Trimethylpyrazine		39,41	41	41	41	
2-Ethylpyrazine	39,41	39,41	41			41
2-Ethyl-5-methylpyrazine				41		
2-Ethyl-6-methylpyrazine	39,41					41
2,6-Diethylpyrazine	39,41					41
2,5-Dimethyl-3-ethylpyrazine	41	39,41	41		41	
2,6-Diethyl-3-methylpyrazine	39	41	41			
Pyrazine, mol wt 136		41				
Pyrazine, mol wt 136		41				
Pyrazine, mol wt 150	41	41	41			41
Pyrazine, mol wt 150	41	41	41			41
Pyrazine, mol wt 164	41	41				41
Pyrazine, mol wt 178		41				
4-methyl-n-propylpyrazine	39	39				
2,5-diketo-3,6-dimethylpiperazine	39					
Pyrrole	39	39				

Table XII. Volatile Compounds Produced from Hydroxy Amino Compounds (cont'd)

Compound	Serine	Threonine	Ethanol-amine	Glucos-amine	4-Amino-3-hydroxy-butyric	Alanyl-serine
2-Methylpyrrole	39	39				
Dimethylpyrrole	39					
3-Methyl-4-ethylpyrrole	39	39				
Ethylamine	39					
Ethanolamine	39					
Ammonia	39					
Acetaldehyde	39	39				
Propionaldehyde	39	39				
Paraldehyde	39					

From references (39) and (41).

tryptophan, lysine, aspartic, asparagine, glutamic, glutamine, adenine, and adenosine were individually subjected to pyrolysis. On the other hand, pyrazines were obtained by heating individual amino-hydroxy compounds, notably those having amino and hydroxy groups in adjacent carbon positions without the participation of sugars. It is obvious that the amino-hydroxy compounds themselves serve as a source of both carbon and nitrogen in the pyrazine molecule. Thus, these compounds are important precursors in foods, e.g., ethanolamine, glucosamine, serine, threonine, 4-amino-3-hydroxybuty ric acid, and alanylserine. The composition of pyrazine mixtures and the relative amounts of each pyrazine produced varied from one amino-hydroxy compound to another. For example, pyrazine, 2-methyl pyrazine, 2,5-dimethyl pyrazine, 2,3-dimethyl pyrazine, and trimethylpyrazine with(2-methyl pyrazine for the largest peak) resulted from glucosamine; while trimethyl-, 2,5-dimethyl-, and 2,5-dimethyl-3-ethyl pyrazine(with 2,5-dimethyl pyrazine for the largest peak were obtained from 4-amino-3-hydroxy-butyric acid (41). Dawes and Edwards (42) speculated on the mechanism of formation of substituted pyrazines in sugar-amine systems.

f- Nonenzymic Browning Reaction. Reactions induced by heating amino acids and sugars are known as nonenzymic browning or Maillard reactions. Meat flavors are also generated in reactions of this type. This had been the subject of several reviews (43, 44, 45, 46). The Maillard reaction is the reaction between an amino compound (amine, amino acid, peptide, or a protein) and a glucosidic hydroxy group in a sugar . Hodge (44) proposed a seven-step mechanism:
1- Sugar-amino acid condensation (formation of N-substituted glycosylamine).
2- Amadori rearrangement (rearrangement to produce a substituted 1-amino-1-deoxy-2-ketose).
3- Sugar dehydration.
4- Sugar fragmentation.
5- Amino acid degradation.
6- Aldol condensation.
7- Aldehyde-amine polymerization.
Products of the Maillard reaction include aliphatic aldehydes, furfural, furfural derivatives, ketone, and 1,2-dicarbonyl compounds. The aliphatic aldehydes are produced by the oxidative degradation of the amino acids (known as Strecker degradation). The resulting aldehdes contain one carbon atom less than the initial amino acid. The Strecker degradation can occur by reaction of amino acids either with Amadori rearrangement products

or with dicarbonyl compounds present in the reaction
mixture. Furfural and its derivatives are produced by
dehydration of the amadori rearrangement products, and
they can participate in the Strecker degradation.

There is no definite relatioship between the odor
produced and the temperature to which various individual
amino acids were heated with glucose (47). For example,
with leucine and threonine the odor was pleasant at 100°,
but unpleasant at 180°C. With histidine an odor appeared at 180°C. Irrespective of the temperature, glycine,
α-alanine, β-alanine and glutamic acid produced only
the odor of burnt sugar when heated with glucose; while
tryptophan, tyrosine, argininine, and α-aminobutyric
acid produced the burnt sugar odor only at 180°C.

Table XIII lists the volatile compounds produced
by the reaction of cysteine and cystine with glucose
and pyruvaldehyde (48).At 160°C, there are not too many
differences between the aroma produced by the reaction
of pyruvaldehyde and glucose. Pyrazines had been identified as the volatiles contributing to the flavors produced by the sugar-amino acid reaction. 2,5-Dimethyl-,
and 2,5-dimethyl-3-ethylpyrazine were detected in the
volatiles resulting from heating cysteine and cystine
with pyruvaldehyde. Trimethylpyrazine was found only in
the presence of cysteine. While methyl-,2,5-dimethyl-,
2-methyl-3-ethyl-, 2-methyl-6-ethyl-, and 2,5-dimethyl-
3-ethylpyrazine were products of heating glucose with
cysteine; methylpyrazine was the only pyrazine present
in   the volatiles resulting from a mixture of glucose
and cystine. Several thiophenes were detected in the
volatiles produced by the reaction of cysteine with
pyruvaldehyde except 2-thiophenoic acid which existed
only in the presence of cystine. Thiophene and hydroxy-
thiophene were found in the volatiles produced from heated glucose-cysteine.Thiophenes generally give a little
unpleasant and characteristic odor. Thiazoles were produced from sulfur-containing amino acids in the presence
of either glucose or pyruvaldehyde, but their species
were different.

g- Peptides. In flavor chemistry, peptides have
been studied as components which contribute a bitter,
sour and sweet taste. Reports on peptides as contrtrib-
utors to discoloration and aroma of foods was seldom
reported.Relating to the production of food flavor and
nonenzymic browning, studies were concentrated on amino
acids, amines and proteins. The order of the browning
rate of amino compounds is : tetraglycine > triglycine >
diglycine > DL-alanyl-DL-alanine > glycine > DL-alanine,
and the reactivity of peptide is much higher than that

Table XIII. Volatile Compounds Produced by the Reaction of Sulfur-Containing Amino Acids with Glucose or Pyruvaldehyde.

Volatile Compounds	Glucose and Cysteine	Cystine	Pyruvaldehyde & Cysteine	Cystine
Thiophene	X	-	X	-
Hydroxy thiophene	X	-	-	-
2-Methylthiophene	-	-	-	-
3-Methyl Thiophene	-	-	X	-
2,5-Dimethylthiophene	-	-	X	-
2,3-Dihydro-4(or 5)-ethyl-thiophene	-	-	-	-
2(or 3-)Ehtylthiophene	-	-	X	-
2-Methyl-3(or 4-)ethyl-Thiophene	-	-	-	-
2,3,5-Trimethylthiophene	-	-	-	-
3-Methyl-n-propylthiophene	-	-	-	-
2,4-Dimethyl-5-ethylthiophene	-	-	-	-
2-Thiophenoic acid	-	-	-	X
2-Methyltetrahydrothiopehen-3-one	-	-	X	-
Thiazole	-	-	X	-
2-Ethylthiazole	-	-	X	-
2-Methylthiazole	-	X	-	X
5(or 4)-Methyl thiazole	-	-	X	-
5(or 4)-Ethylthiazole	-	-	X	-
Trimethyl thiazole	-	X	-	-
2-Methylthiazoline	X	-	-	-
2-Acetyl-4-methylthiazole	-	-	-	X
Pyridine	X	X	-	-
α-Picoline	X	-	-	-
β-Picoline	X	X	-	-
2-Methyl-5-ethylpyridine	-	X	-	-
Methylpyrazine	X	X	-	-
2,5-Dimethylpyrazine	X	-	X	X
2-Methyl-6-ethylpyrazine	X	-	-	-
2-Methyl-3-ethylpyrazine	X	-	-	-
2,5-Dimethyl-3-ethylpyrazine	X	-	X	X
Trimethylpyrazine	-	-	X	-
Furfural	-	X	-	-
2-Methyl-5-ethylfuran	-	X	-	-
2-Acetylfuran	X	-	-	-
2-Acetyl-5-methylfuran	X	-	-	-
Furfural alcohol	-	X	-	-
5-Methylfurfural	-	X	-	X
2,5-Dimethyl-3-ethylfuran	-	-	-	X
2-Furoic acid	-	-	X	X
Phenol	-	-	X	-
p-Methylbenzoic Acid	-	-	-	X

Table XIII.   Volatile Compounds Produced by the Reaction of
Sulfur-Containing Amino Acids with Glucose or
Pyruvaldehyde. (cont'd)

Volatile Compounds	Glucose and Cysteine	Cystine	Pyruvaldehyde & Cysteine	Cystine
Ethyl Alcohol	X	X	-	-
2,4(or 5)Dimethylthiazole	-	X	-	-
2-Methyl-4(or 5) ethylthiazole	-	X	-	X
2-Ethylthiazoline	-	-	X	-
Benzoic Acid	-	-	X	X
p-Methylbenzoic Acid	-	-	X	X
Acetic Acid	X	X	-	X
2-Methylthiazoline	-	X	-	-
Furfuryl alcohol	-	X	-	-
Benzene	-	-	X	-

From reference (48)

of the amino acid(49). The taste of glycyl-L-leucine is
very bitter, but the product of its reaction with glyoxal at 100°C,pH 5.00; has an astringent, a little sour
and later a mild taste. A series of new pyrazinones
2-(3'-alkyl-2'-oxo-pyrazin-1'-yl)alkyl acids were prepared from various dipeptides, Table XIV. Pyrazine products resulting from the reaction of amino acids with
carbonyl compounds and by the pyrolysis of some amino
acids were considered to be stable, especially when its
substituents are alkyl groups. Pyrazines with substituents other than alkyl group such as in the case of
dihydropyrazine was shown to be reactive. The addition
of pyrazinone to dipeptide-glyoxal solution increased
the browning while amino acid-pyrazinone or glyoxal-
pyrazinone solution caused no browning (49). One can
assume that pyrazinone plays a role in the browning reaction, and it probably reacts with neither the amino
acid nor with glyoxal, but with the product produced in
the reaction. The experimental results indicated that
2-(3'-methyl-2'-oxo-pyrazin-1'-yl)propionic acid was
more active in browning than 2-(2'-oxopyrazin-1'-yl)
isocaproic acid, the substituent may be considered as a
causative factor.

In a study on lipid browning,Dugan and Rao (50)
reported that the reactions proceed readily at low moisture levels(2.5%) and ambient or elevated temperatures.
The reaction between phosphatidyl ethanolamine(PE) and
nonanal (Table XV) on protein matrix (lipid-free beef
muscle fibers) indicated that the carbonyl reactions
proceeded with both amino groups from PE and from the
protein matrix. Certain amino acids, such as, lysine,
alanine, phenylalanine, and tyrosine reacted with aldehydes (nonanals or oxidation products of unsaturated
PE), sufficiently to reduce their quantity in the total
amino acid content in the system. The reactions were
competitive that the presence of PE had a sparing effect on amino acids in the protein matrix. This observation may provide a rationale for the changes in texture,color, flavor, and nutritive value of dried foods
containing proteins and phospholipids.

Furans in Meat

Tonsbeek et al. (51) reported the presence of
4-hydroxy-2,5-dimethyl-3(2H)-furanone and 4-hydroxy-5-
methyl-3(2H)-furanone in beef broth. The precursors of
these compounds are ribose-5-phosphate and either taurine or pyrrolidone carboxylic acid. In the products of
the reaction of these two hydroxy furanones with hydrogen sulfide,were reported substituted 3-mercaptofurans,

Table XIV. Pyrazinones From Various Dipeptides In The Reaction with With Glyoxal

Name of Pyrazinone Formed	Reacting Dipeptides
2-(2'-Oxo-pyraxin-1'-yl)acetic acid	Glycylglycine
2-(2'-oxo-pyrazin-1'-yl)propionic acid	Glycyl-DL-alanine
2-(3'-Methyl-2'oxo-pyrazin-1'-yl)propionic acid	DL-Alanyl-DL-alanine
2-(3'-Methyl-2'oxo-pyrazin-1'yl)isovaleric acid	DL-Alanyl-DL-valine
2-(3'-Methyl-2'-oxo-pyrazin-1'-yl)valeric acid	DL-Alanyl-DL-norvaline
2-(3'-Methyl-2'-oxo-pyrazin-1'-yl)caproic	DL-Alanyl-DL-norleucine
2-(3'-Methyl-2'-oxo-pyrazin-1'-yl)isocaproic	DL-Analym-DL-leucine
2-(3'-Isobutyl-2'-oxo-pyrazin-1'-yl)acetic	DL-Leucylglycine
2-(3'-Isobutyl-2'-oxo-pyrazin-1'-yl)isocaproic	DL-Leucyl-DL-leucine

From reference (49)

Table XV. Amino Acid Analysis of Muscle Fiber after Storage with Phospholipids and/or Aldehydes

Amino Acid	Gram Residue / 100 gram of Sample					
	Original Muscle fiber before Browning Reaction	Muscle fiber plus hydrogenated PE	Muscle fiber plus PE	Muscle fiber plus Phosphatidal Ethanolamine	Muscle fiber plus PE and Monanal	Muscle fiber plus Nonanal
Lysine	1.74	1.77	1.21	1.00	0.71	0.39
Histidine	0.55	0.61	0.99	0.65	0.79	0.74
Arginine	1.24	1.20	1.11	1.02	1.09	1.20
Aspartic Acid	1.81	1.80	1.58	1.65	1.49	1.74
Threonine	1.08	1.05	1.00	0.89	1.10	0.89
Serine	0.68	0.71	0.76	0.68	0.71	0.67
Glutamic Acid	2.74	2.76	.267	2.57	2.69	2.95
Proline	0.69	0.63	1.15	0.72	1.00	0.59
Glycine	0.72	0.69	0.92	0.89	0.90	0.86
Alanine	1.04	1.12	0.96	0.82	0.76	0.70
Cysteine	-	-	-	-	-	-
Valine	1.06	1.11	0.97	1.21	1.21	0.95
Methionine	0.23	0.25	0.00	0.05	0.00	0.00
Isoleucine	1.03	1.00	1.10	0.98	1.02	0.03
Leucine	1.55	1.49	1.51	1.62	1.58	1.55
Tyrosine	0.62	0.65	0.49	0.42	0.27	0.10
Phenylalanine	0.84	0.92	0.62	0.27	0.32	0.10

From reference (50)

and 3-mercaptothiophenes in various degrees of saturat-
ion(52).The thiophene analogs of the hydroxyfuranones
starting materials were also formed in these reactions.
These analogs in turn afforded still more mercaptothio-
phenes and mercaptothiophenones.The reaction systems
and many of purified isolated compounds had meat-like
odors. As hydroxyfuranones and hydrogen sulfide had
been reported in boiled beef(51), it is expected that
their reaction products are also present, though in concentrations undetected by the available instrumental
techniques.

Taste of Flavor Precursors

The overall "flavor" sensation may conveniently be divided into sensation due to "taste" and "aroma". In the
case of cooked meats, the contribution of aromatic compounds would be large in comparison with that due to
the non-volatile taste-bearing components. All D-amino
acids have sweet taste. With the exception of L-phenylalanine which is bitter,and l-alanine which is slightly
sweet, all L-amino acids are tasteless. Free L-glutamic
acid has a well known brothy taste. Some of the L-glut-
amyl dipeptides formed from coupling an amino acid to
the α-carboxyl group of L-glutamic acid were found to
taste brothy: glutamylaspartic , glutamylthreonine ,
glutamylserine, and glutamylglutamic (53). Glutamylglycylserine is responsible for the brothy taste of an
enzymatically modified soybean protein. About thirty
acidic oligopeptides were isolated and identified in a
flavor potentiating fraction from fish protein hydrolysate (54). Four dipeptides (glutamylaspartic, glutamylglutamic, glutamylserine, and threonylglutamic), and
five tripeptides (aspartylglutamylserine,glutamylaspar-
tylglutamic, glutamylglutamineglutamic, glutamylglycylserine, and serylglutamylglutamic) had a flavor quantitatively resembling that of monosodium glutamate(MSG).
The threshold level of glutamylglycylserine, for example, was estimated to be approximately 0.2% in water at
pH 5.0 , whereas that of MSG was almost one tenth of
this level under the same conditions. Some isolated
peptides were reported to exhibit bitter taste(glutamyl-
aspartylvaline, aspartylleucine, isoleucylglutamyl-
glutamic, and isoleucylglutamic) while others had a
flat taste. A strong synergistic relationship exists
between nucleotides and MSG with the result that blends
of MSG and nucleotides exceed the potency and versatil-
ity of either material alone.The optimum blend which
offers the greatest effectiveness for the least cost is
a blend of 95% MSG and 5% nucleotides. Desirable fla-

vors were enhanced by addition of sodium inosinate re-
gardless of the type of the meat or cooking method (55).
Disodium inosinate consistently produced an impression
of greater viscosity and increased flavor.

Amino acids play an important part in the palatable taste of foods. When glycine was added to a 1% NaCl
solution containing L-glutamic or L-aspartic and IMP
plus GMP at 0.1% , the palatable taste of the medium
was significantly improved. The ternary synergism among
L-glutamic or L-aspartic,5'-nucleotides and glycine is
different from the binary synergism already known between L-glutamic acid and 5'-nucleotides (56). The ternary synergism is significant at the concentration of
the stimulus   threshold   of these components. The
ternary synergism was not due to the sweet taste of
glycine. No binary synergism of palatable taste was observed between glycine and L-glutamic or L-aspartic, or
glycine and 5'-nucleotides.Extending the study to include other α-amino acids,indicated the absemce of any
binary synergism of palatable taste between IMP plus
GMP or MSG and glycine, L-alanine, L-serine, L-histi-
dine-HCl, L-methionine, or DL-tryptophan in 1% NaCl so-
lution(57).However, L-alanine(0.05), L-cystine(0.05),
glycine(0.10), L-histidine-HCl(0.004), L-methionine
(0.03), L-proline(0.20), L-serine(0.10), DL-tryptophan
(0.015), or L-valine(0.15g./dl.) gave a ternary synergism with IMP plus GMP and MSG(0.01%, respectively) in
1% NaCl solution. When two or three of this group of
amino acids were added to 1% NaCl solution containing
MSG and IMP plus GMP at 0.01%, the ternary synergism
caused by these amino acids was estimated as the alge-
braic sum of the activity of the individual amino acids
(58).

Pharmacological Effects

Red meat flavor precursors have well known functions in
human nutrition. The evidence of their pharmacological
properties has been known for many decades before their
contributions to flavor were established. Such properties of amino acids, peptides, sugars,nucleotides..etc
have been reviewed in human physiology ,human biochemistry and human nutrition books and reviews (59,60).

Meat extract, which is a concentrate of aqueous
beef extract, is claimed to be a stimulant. Products
containing meat extract, have been regarded as general
tonics and stimulants, assisting recovery from exhaus-
tion and fatigue. Precise information about these effects are lacking. It is, of course possible that the
well-known promotion of gastric and intestinal secre-

tion and motility might cause secondary,more generalized effects and result in a real or apparent stimulation. Injecting creatine in frogs before setting them
to work on a treadmill, proved to be responsible for
improving muscular performance and hastened recovery
after work(61). In his classical fistulas experiments
with dogs, Pavlov (62) had indicated that meat extract
was a powerful stimulant of gastric secretion. He evaluated some of meat extract components and concluded
that creatine,hypoxanthine, xanthine,leucine and a mixture of inosine and hypoxanthine were ineffective. Later,Krimberg and Komarov (63,64,65) and Korchow (66)
demonstrated that carnosine at a concentration of 0.02
g./kg was the compound responsible for the gastric
stimulating properties. Using a dose of 0.005 g./kg. ,
Schwarz and Goldschmidt reported no stimulation of secretion following the administration of carnosine into
dogs with gastric fistulas(67). Evaluating pure samples
of carnosine, carnitine and methyl guanidine for their
effect as stimulant of gastric secretion, proved that
carnosine was the most powerful one. Intravenous injection of carnosine was most effective than subcutaneous
and oral administration. The principal function of anserine and carnosine involves the coupling of phospho-
rylation with glycolysis and the synthesis of adenosine
triphosphate. Carnitine which had been reported in meat
extracts is identical with vitamin B_7 (68). Methionine
has been reported to be beneficial in the prevention
and treatment of liver injury due to poisoning by arse-
nic, chloroform, carbon tetrachoride, or trinitro -
toluene.It has been recommended in the treatment of
eclampsia, shock, infectious hepatitis and cirhosis of
the liver. It has been prescribed in the management of
obese, and patients with severe burns(69).

    "The whole field of flavor precursor chemistry is
in its infancy, and it is reasonable to say that no
study — be it academic or commercial — of a natural
flavor can be considered complete unless it includes
the precursors of that flavor. An approach to any flavor problem which recognizes the importance of the two
complementary approaches is more likely to give a complete picture of the flavor and the mechanism of its
formation than attempts to interpret its chemical
constitution", Rohan's statement(70) expresses my views.

ABSTRACT
    Red meats exhibit distinct taste and flavor which
are characteristic of their animal speies. Age of the
animal, type of feed, animal condition, type of meat

cuts, meat processing, etc.,strongly affect the taste
and flavor of individual meats. Non-volatile nitrogen
and sulfur compounds play an important role in the
development of the characteristic flavors of red meats.
Their pharmacological effect, relative importance, and
their role in the formation of flavor components were
discussed. Speculations on the reasons for differences
in red meat flavors th$_a$t intrigue flavor chemists were
reported.

Literature Cited

1. Wong, E., Nixon, L.N., Johnson, C.B., J. Agr. Food
   Chem. (1975), 23, 495.
2. Patterson, R.L.S., J. Sci. Food Agr. (1968), 19, 31.
3. Crocker, E.C., Food Res. (1948), 13, 179.
4. Bender, A.E., Wood, T., Palgrave,J.A., J. Sci.
   Food Agr. (1958), 9, 812.
5. Wood, T., J. Sci. Food Agr. (1961), 12, 61.
6. Batzer,O.F., Santoro, A.T., Landmann, W.A., J. Agr.
   Food Chem. (1962), 10, 94.
7. Batzer, O.F., Santoro, A.T., Tan, M.C., Landmann,
   W.A., Schweigert, B.S., J. Agr. Food Chem. (1960),
   8, 498.
8. Hornstein, J., "Chemistry and Physiology of Flavors",Schultz, H.W., Day, E.A., Libbey, L.M., Eds.
   pp. 228-250, Avi Publishing Co., Westport, Conn.
   1967.
9. Macy, R.L., Jr., Naumann, H.D., Bailey, M.E., J.
   Food Sci. (1964), 29, 136.
10. Macy, R.L., Jr., Naumann, H.D., Bailey, M.E., J.
    Food Sci. (1964), 29, 142.
11. Wasserman, A.E., Gray, N., J. Food Sci. (1965), 30,
    801.
12. Zaika, L.L., Wasserman, A.E., Monk, C.R., Jr.,
    Salay, J., J. Food Sci. (1968), 33, 53.
13. Mabrouk, A.F., Jarboe, J.K., O'Connor, E.M., J. Agr.
    Food Chem. (1969), 17, 5.
14. Mabrouk, A.F., J. Agr. Food Chem. (1973), 21, 942.
15. Mabrouk, A.F., Kramer, R., Jarboe, J.K., Alabran,
    D.M., Unpublished data.
16. Paulson, J.C., Deatherage, F.E., Almy, E.F., J. Am.
    Chem. Soc. (1953), 75, 2039.
17. Jarboe, J.K., Mabrouk, A.F., J. Agr. Food Chem.
    (1974), 22, 787.
18. Dvorak, Z., Vognarova, I., J. Sci. Food Agr. (1969),
    20, 146.
19. Johnson, A.R., Vickery, J.R., J. Sci. Food Agr.
    (1964), 15, 695.

20. Dahl, O., J. Agr. Food Chem. (1963), 11, 350.
21. Bendall, J.R., J. Soc. Chem. Ind. (1946), T226.
22. Nakajima, N., Ichikawa, K., Kamada, M., Fujita, E., J. Agr. Chem. Soc. Japan (1961), 35, 797.
23. Mabrouk, A.F., Abstracts of Papers. 86, Division of Agricultural And Food Chemistry, 160th ACS National Meeting, Chicago, Illinois, September 1970.
24. Mabrouk, A.F., Abstracts of Papers. 39, Division of Agricultural And Food Chemistry, 162nd ACS National Meeting, Washington, D.C., September 1971.
25 Terasaka, M., Takeda Scientific Information Bulletin No. 26.
26. Fujimaki, M., Chuyen, N. V., Matsumoto, T., Kurata, T., J. Agr. Chem. Soc. Japan (1970), 44, 275.
27. Macy, R.L., Jr., Naumann, H.D., Bailey, M.E., J. Food Sci. (1970), 35, 78.
28. Macy, R.L., Jr., Naumann, H.D., Bailey, M.E., J. Food Sci.(1970), 35, 81.
29. Macy, R.L., Jr., Naumann, H.D., Bailey, M.E., J. Food Sci. (1970), 35, 83.
30. Craske, J.D., Reuter, F.H., J. Sci Food Agr. (1965), 16, 243.
31. Miyake, M., Tanaka, A., J.Food Sci. (1971), 36, 674.
32. Merritt, C., Jr., Robertson, D.H., J.Gas Chromatog. (1967), 4, 96.
33. Giacabbo, H., Simon, W., Pharm. Acta Helv. (1964), 39, 162.
34. Vollmin, J., Kriemler, P., Omura, J., Seible, J., Simon, W., Microchem. J. (1966), 11, 73.
35. Fujimaki, M., Kato, S., Kurata, T., Agr. Biol. Chem. (1969), 33, 1144.
36. Kato, S., Kurata, T., Fujimaki, M., Agr. Biol. Chem. (1973), 37, 1759.
37. Tondeur, R., Sion, R., Deray, E., Bull. Soc. Chim. France (1964), 2493.
38. Kato, S., Kurata, T., Fujimaki, M., Agr. Biol. Chem. (1971), 35, 2106.
39. Kato, S., Kurata, T., Ishitsuka, R., Fujimaki,M., Agr. Biol. Chem. (1970), 34, 1826.
40. Maga, J.A., Sizer, C.E., CRC Critical Rev. Food Technol., (1973), 4, 39.
41. Wang, P., Odell, G.V., J. Agr. Food Chem.(1973), 21,868.
42. Dawes, I.W., Edwards, R.A., Chem. Ind. (London), (1966), 2203.
43. Hodge, J.E., J. Agr. Food Chem. (1953), 1, 928.
44. Hodge, J.E., "The Chemistry and Physiology of Flavors", Schultz, H.W., Day, E.A., Libbey, L.N., Eds., pp 465-491, Avi Publishing Co., Westport, Conn. , 1967.

45. Reynolds, T.M., Advan. Food Res. (1965), 14, 167.
46. Reynolds, T.M., Advan Food Res. (1963), 12, 1.
47. Herz, W.J., Shallenberger, R.S., Food Res. (1960), 25, 491.
48. Kato, S., Kurata, T., Fujimaki, M., Agr. Biol. Chem. (1973), 37,539.
49. Chuyen,N.V., Kurata, T., Fujimaki, M., Agr. Biol. Chem. (1973), 37, 327.
50. Dugan, R.L., Jr., Rao, G.V. "Evaluation of The Flavor Contribution of Products of the Maillard Reaction ". Technical Report 72-27F, p 110,US Army Natick Development Center, Natick, MA ,1972.
51. Tonsbeek, C.H.T., Blancken, A.J., van de Weerdhof, T., J. Agr. Food Chem. (1968), 16,1016.
52. van den Ouweland, G.A.M., Peer, H.G.(to Lever Brothers Co.),U.S. Patent 3,697,295(July 1,1968).
53. Arai, S., Yamashita, M., Noguchi, M., Fujimaki, M., Agr. Biol.Chem. (1973), 37, 151.
54. Noguchi, M., Arai, S., Yamashita, M., Kato, H., Fujimaki, M., J.Agr. Food Chem. (1975), 23, 49.
55. Kuninaka, A.,"The Chemistry and Physiology of Flavors", Schultz, H.W., Day, E.A., Libbey, L.N., Eds., pp 515-535, Avi Publishing Co., Westport, Conn., 1967.
56. Yokotsuka, T., Saito, N., Okuhara, A., Tanaka, T., Nippon Nogei Kagaku Kaishi (1969), 43, 165.
57. Tanaka, T.,Saito, N.,Okuhara,A.,Yokotsuka, T., Nippon Nogei Kagaku Kaishi (1969), 43,171.
58. Tanaka, T., Saito, N., Okuhara, A., Yokotsuka, T., Nippon, Nogei Kagaku Kasshi (1969), 43, 263.
59. Brobeck, J.R., Ed. " Best & Taylor's Physiological Basis of Medical Practice", The Williams & Wilkins Company, Baltimore,1973
60. Babkin, B.P. "Secretory Mechanism of the Digestive Glands" 2nd Ed. ,p 1027, Paul B.Hoeber, Inc.,New York, 1950.
61. Kobart, E.H., Arch. Exp.Path. Pharmak. (1882-3),15, 56.
62. Pavlov, I.P."The Work of the Digestive Glands",2nd English Ed. by W.H.Thompson,C.Griffin & Co.,1910.
63. Krimberg, R., Komarov, S.A., Biochem. Z.(1926),171, 169. Thru CA,20,3313 (1926).
64. Krimberg, R., Komarov, S.A.,Biochem.Z. (1927),184, 442. Thru CA,21,2309 (1927).
65. Krimberg, R., Komarov, S.A., Biochem. Z.(1928),194, 410. Thru CA,22,2779 (1928).
66. Korchow, A. Biochem.Z.(1927),190, 188. Thru CA,22, 1625 (1928).

67. Schwarz, C., Goldschmidt, E., Arch. ges. Physiol.
    (Pfluger's) (1924), <u>202</u>, 435. Thru CA,18, 2027,
    (1924).
68. Carter, H.E., Bhattacharyya, P.K., Weidman, K.R.,
    Fraenkel,G., Arch. Biochem. Biophys. (1952),<u>38</u>,405.
69. Osol, A.E., Pratt, R., Altschule, M.D., " The
    United States Dispensatory And Physicians' Phar-
    macology",pp 715-716, J.P.Lippincott Company,
    Philadelphia & toronto,1972.
70. Rohan, T.A., The Flavour Industry (1971), <u>2</u>, 147.

11

Furans Substituted at the Three Position with Sulfur

WILLIAM J. EVERS, HOWARD H. HEINSOHN, JR.,
BERNARD J. MAYERS, and ANNE SANDERSON

International Flavors and Fragrances, Inc., 1515 Highway 36, Union Beach, N. J. 07735

During recent years the analysis of cooked meats and of model systems has been reported by numerous investigators; leading references include Wilson et al (1), Mussinan and Katz (2), Persson and von Sydow (3), Schutte (4), and van den Ouweland and Peer (5).

We now report the isolation, identification and synthesis of two important meat flavor and aroma chemicals, 2-methyl-3-furanthiol (1) and bis (2-methyl-3-furyl) disulfide (2).

      1   2

While preparing small batches (1000 g) of a processed meat flavor which consisted of a mixture of L-cysteine hydrochloride, thiamine chloride hydrochloride, hydrolyzed vegetable protein and water heated at reflux for four hours (6), it was noted that the condensate possessed an intense roasted aroma. The material responsible for the aroma proved to be extractable with methylene chloride but when the methylene chloride was removed, only a minute amount of material remained. In order to obtain workable quantities of aroma material for analysis a large scale preparation was undertaken. This reaction totaled 4,000 lbs. and gave 40 gallons condensate from which 16 liters of methylene chloride extract was obtained. Careful solvent removal of the extract under mild vacuum and final concentration with a stream of nitrogen gave

50 ml of concentrate possessing a powerful roast aroma.
Thin-layer chromatography of the concentrate indicated the presence of a readily isolable compound
near the solvent front. Repeated preparative thin-
layer chromatography gave, from 2.4 g of concentrate,
a 66 mg sample of the disulfide 2. The structure is
assigned on the basis of mass spectral, proton magnetic
resonance and infrared data. When the concentrate was
subjected to examination by GLC, it gave a succession
of peaks blending into a complex chromotogram. One
particular section of the GLC effluent when organoleptically evaluated had what was described as a "pot
roast" type aroma. Isolation of the compound or compounds responsible for this odor was undertaken. As a
result of repeated trapping and rechromatography
it became apparent that the pot roast compound was
present in very minute amounts. With much trial and
error, a scheme for the isolation of this compound was
worked out. Thus, from 4 ml of concentrate we obtained
2-3 λ of the furanthiol 1. As in the case of the disulfide, the structure proposed was based on mass
spectral, proton magnetic resonance and infrared data.
In order to confirm the proposed structures of 1
and 2, their synthesis was undertaken and is outlined
in Table I. Sulfonation of 5-methyl-2-furoic acid
(3) with fuming sulfuric acid gave the known 2-methyl-
3-sulfo-5-furoic acid (4) (7) which was decarboxylated
with mercuric chloride to yield 2-methyl-3-furansul-
fonic acid (5). Treatment of 5 with thionyl chloride
gave the sulfonyl chloride 6, which was reduced with
lithium aluminum hydride, giving 2-methyl-3-furanthiol
(1). This material has an identical mass spectrum,
proton magnetic resonance and infrared spectrum as the
materials isolated from the reaction meat flavor. Oxidation of 1 with iodine gave bis (2-methyl-3-furyl)
disulfide (2) which also has identical spectral properties as the isolated material. These syntheses
thus confirm the structures proposed from the analytical data for the isolated materials (8).
Similar compounds were synthesized in order to
compare the flavor properties of these compounds. The
corresponding 2,5-dimethyl-3-furanthiol (8) was prepared by hydrolysis of 2,5-dimethyl-3-thioacetylfuran
(7)(10).
Treatment of 2,5-dimethyl furan with either sulfur monochloride or sulfur dichloride gave, in low
yield, a mixture containing both bis (2,5-dimethyl-3-
furyl) sulfide (10c) and bis (2,5-dimethyl-3-furyl) disulfide (10d). The corresponding tri- and tetrasulfides of 2 were also prepared. Thus the reaction of 1

TABLE I

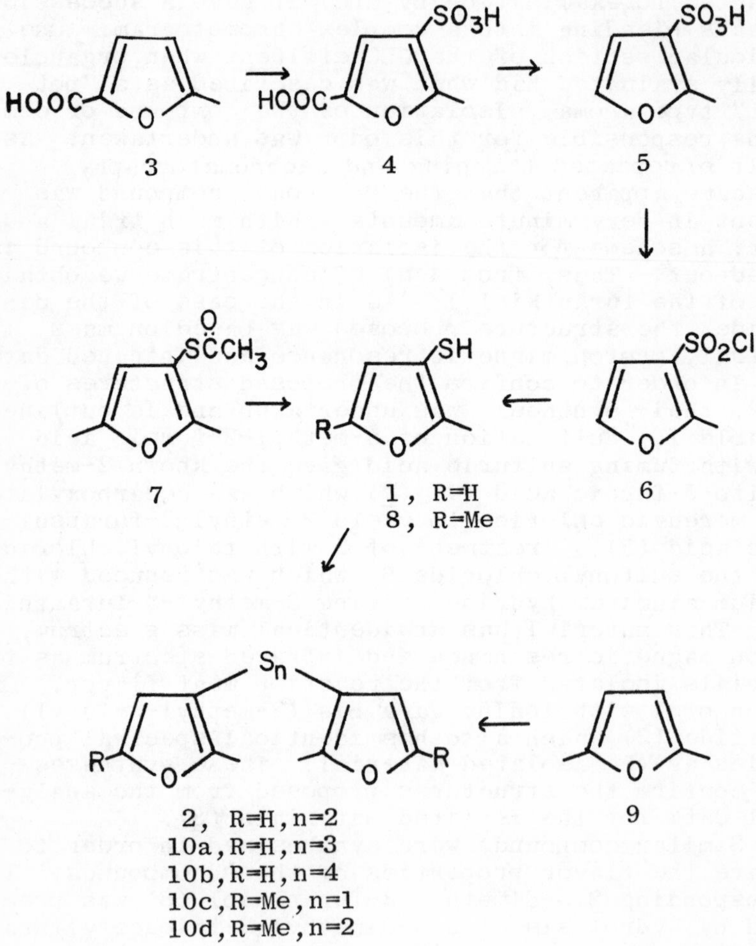
SO3H
HOOC
O
3
SO3H
HOOC
O
4
SO3H
O
5
SO2Cl
O
6
O
SCCH3
O
7
SH
R
O
1, R=H
8, R=Me
Sn
R
O
O
R
2, R=H, n=2
10a, R=H, n=3
10b, R=H, n=4
10c, R=Me, n=1
10d, R=Me, n=2
O
9

with sulfur dichloride gave bis (2-methyl-3-furyl) trisulfide (10a) and the reaction of 1 with sulfur monochloride gave bis (2-methyl-3-furyl) tetrasulfide (10b). For comparison, 5-methyl-2-furanthiol (12) and bis (5-methyl-2-furyl) disulfide (13) were prepared. Treatment of the lithio derivative of sylvan (11) with sulfur gave the thiol 12 which was oxidized with iodine to give 13.

11   12   13

A summary of the organoleptic evaluations of the various compounds is contained in Table II. It is apparent from a brief study of the table that all of the furans where the sulfur atom is bound to the ß carbon have meaty aroma and taste characteristics. However, when the sulfur atom is in the α position only hydrogen sulfide like, burnt and chemical notes are observed. It is somewhat surprising that a one carbon shift in substitution would produce this substantial change in flavor character.

Experimental

Isolation of 1 and 2. In a suitable vessel a mixture of 35 lbs. of thiamine chloride hydrochloride, 35 lbs. of L-cysteine hydrochloride, 1,238 lbs. of vegetable protein hydrolysate (carbohydrate free) and 2,692 lbs. of water was heated at reflux for four hours. After 45 minutes, 40 gallons of condensate was removed during $3\frac{1}{4}$ hrs. Each gallon was extracted with 400 ml of CH_2Cl_2 and the extracts were combined and dried with sodium sulfate. Solvent removal in mild (100 mm) vacuum and final concentration with a stream of dry nitrogen gave 50 ml of concentrate.

2-Methyl-3-furanthiol (1): Preparative GLC of the concentrate (2 x 2 ml) on an 8' x 3/4", 25% SE 30 on A/W DMCS column, programmed from 70° to 225°C at 1°/min yields about 300 λ containing a strong pot roast aroma. Rechromatography of the 300 λ on a 10' x 3/8", 20% CBW 20M A/W DMCS column programmed from 80° to 225°C at 1°/min yields about 100 λ of material containing a very intense pot roast aroma. Final chromatography of this material on a 10' x 1/4" glass, 20% Apiezon L column gave 2-3 λ of 2-methyl-3-furanthiol

TABLE II

Flavor Characteristics

Compound	Medium	Description
2-Methyl-3-furanthiol (1)	Water	Sweet meat, beef broth, HVP like. Roasted meat aroma.
2,5-Dimethyl-3-furan-thiol (8)	Water	Strong meaty taste. Roasted meat aroma.
5-Methyl-2-furanthiol (12)	Water	Hydrogen sulfide like, burnt, sulfury-not meaty.
Bis-(2-methyl-3-furyl) disulfide (2)	Bouillon 0.05 ppm	Meat fullness. Cooked meat aroma.
Bis-(2,5-dimethyl-3-furyl) disulfide (10d)	Bouillon 0.05 ppm	Meaty taste and aroma.
Bis-(2,5-dimethyl-3-furyl) sulfide (10c)	Bouillon 0.05 ppm	Bloody aroma. Boiled meat taste.
Bis-(5-methyl-2-furyl) disulfide (13)	Bouillon 0.05 ppm	Chemical, rubbery aroma and taste.
Bis-(2-methyl-3-furyl) trisulfide (10a)	Water 1 ppm	Brothy
Bis-(2-methyl-3-furyl) tetrasulfide (10b)	Water 10 ppb	Braised beef aroma and taste.

having: ir (neat, KBr plates) 3120, 2960, 2925, 2540
(-$\underline{SH}$), 1595, 1514, 1222, 1136, 1092, 938, 888 and 730
cm^{-1}; nmr (CCl$_4$) δ 2.32 (s,3, α ring C$\underline{H}_3$), 2.47 (s,1
β S$\underline{H}$), 6.19 (d,1,J=1.5Hz, β ring $\underline{H}$), 7.15 ppm (d,1,
J=1.5Hz, α ring $\underline{H}$); mass spectrum (70 ev) m/e (rel.
intensity) 116 (5), 115 (8), 114 (100), 113 (29), 86
(10), 85 (23), 71 (14).

bis (2-Methyl-3-furyl) disulfide (2): Repeated
preparative thin-layer chromotography on 8"x 8" plates
coated with 1.25 mm of silica-gel G (200 λ portions per
plate, developed with 1:10 ether/hexane) gave from 2.4
g of concentrate 0.066 g of bis (2-methyl-3-furyl) disulfide having: ir (neat, KBr plates) 3120, 2955, 2920,
1580, 1512, 1222, 1120, 1085, 936, 885, 730 cm^{-1}; nmr
(CCl$_4$) δ 2.08 (s,3, α ring C$\underline{H}_3$), 6.27 (d,1,J=1.5Hz,
β ring $\underline{H}$), and 7.16 ppm (d,1,J=1.5 Hz, α ring $\underline{H}$); mass
spectrum (70 ev) m/e (rel intensity) 228 (6), 227 (9),
226 (60), 115 (5), 114 (17), 113 (100), 85 (13).

2-Methyl-3-sulfo-5-furoic acid (4): The barium
salt of 4a was prepared in 93% yield from 3 according
to Scully and Brown (7). The free acid was isolated
for nmr by treating the barium salt with H$_2$SO$_4$ followed
by filtration and evaporation under vacuum. Nmr (D$_2$O)
δ 2.62 (s,3, α ring C$\underline{H}_3$) and 7.47 ppm (s,1, β ring $\underline{H}$).

2-Methyl-3-furansulfonic acid (5): A solution of
33 g of the barium salt of 4a in 450 ml of H$_2$O was
treated with 20% H$_2$SO$_4$ until no further precipitation
was observed. After filtration, the filtrate was adjusted to pH 7 with solid NaHCO$_3$. After addition of
26.3 g of mercuric chloride the resulting mixture was
heated at reflux and monitored for CO$_2$ evolution.
When CO$_2$ evolution had stopped, the mixture was cooled
to room temperature and treated with hydrogen sulfide.
After standing overnight precipitated mercuric sulfide
was removed by decantation and filtration. The solution was evaporated to dryness in vacuo and the residue recrystallized from a minimum of boiling water to
give 4.25 g of 2-methyl-3-furan sulfonic acid as the
sodium salt. Ir (nujol mull) 3140, 1240, 1229, 1190,
1110, 1075, 1018, 898 cm^{-1}; nmr (D$_2$O) δ 2.48 (s,3, α
ring C$\underline{H}_3$), 6.62 (d,1,J=2Hz, β ring $\underline{H}$) and 7.43 ppm
(d,1,J=2Hz, α ring $\underline{H}$).

2-Methyl-3-furansulfonyl chloride (6): A mixture of 80 g of 5 (sodium salt), 880 g thionyl chloride and 15 drops of dimethyl formamide was heated at
reflux for 100 min, cooled to room temperature and
filtered. After washing the filter cake with benzene
(20 ml) and combining with the filtrate, concentration
in vacuo gave 58 g of a brown oil. Distillation of
the oil gave 45.4 g of 2-methyl-3-furansufonyl chloride as a light yellow liquid boiling at 55°C at 0.55

torr. and having: ir (neat, KBr plates) 3115, 1570, 1520, 1380, 1360, 1220, 1158, 1129, 1092, 1024, 948, 880, 740 cm^{-1}: nmr (CCl$_4$) δ 2.71 (s,3, α ring CH$_3$), 6.86 (d,1,J=2Hz, β ring $\underline{H}$) and 7.52 ppm (d,1,J=2Hz, α ring $\underline{H}$); mass spectrum (70 ev) m/e (decreasing intensity, mol. ion) 145, 97, 53, 43, 180, and intensity of P+2 (182) is 43% of P.

2-Methyl-3-furanthiol (1): A solution of 45.4 g of 2-methyl-3-furansulfonyl chloride (6) in 540 ml of anhydrous ether was added dropwise to a solution of lithium aluminium hydride in 1200 ml ether. After heating at reflux one hour, excess hydride was destroyed by the addition of ethyl acetate. The resulting mixture was poured into 1,500 ml water and the ether layer separated. The aqueous phase was extracted with ether (4 x 250 ml) and the ether extracts were combined, washed with 5% NaHCO$_3$ solution (150 ml), water (2 x 200 ml) and dried over sodium sulfate. Solvent removal in vacuo gave 18 g crude 1. Distillation gave 11.6 g of 2-methyl-3-furanthiol boiling at 57-58.5° at 36 torr and having: ir, (neat, KBr plates) 3110, 2955, 2925, 2540 (SH), 1594, 1512, 1223, 1134, 1092, 937, 889, 732 cm^{-1}; nmr (CCl$_4$) δ 2.34 (s,3, α ring CH$_3$), 2.47 (s,1, β ring S$\underline{H}$), 6.19 (d,1,J=1.5 Hz, β ring $\underline{H}$) and 7.15 ppm (d,1,J=1.5 Hz, α ring H); mass spectrum (70 eu) m/e (rel. intensity) 116 (5), 115 (8), 114 (100), 113 (29), 86 (10), 85 (23), 71 (13).

bis (2-Methyl-3-furyl) disulfide (2): A solution of 7.4 g of sodium hydroxide, 10 g of sodium carbonate and 21 g of 2-methyl-3-furanthiol (1) in 344 ml of water was prepared. To this solution was added a solution of 77.3 g of potassium iodide and 23.2 g of iodine in 780 ml of water until the iodine color persisted for 1 min. After rendering the solution colorless with 0.1 sodium thiosulfate solution, the mixture was extracted with pentane (3 x 150 ml). The pentane extracts were combined, washed with water (2 x 100 ml) and dried over sodium sulfate. Solvent removal in vacuo gave 10 g of an amber oil. Distillation of the oil gave 15.4 g of bis (2-methyl-3-furyl) disulfide (2) boiling at 77-78° at 0.3 torr and having: ir (neat, KBr plates) 3120, 2955, 2920, 1580, 1512, 1222, 1122, 1082, 938, 885, 730 cm^{-1}; nmr (CCl$_4$) δ 2.07 (s,3, α ring CH$_3$), 6.25 (d,1,J=1.5 Hz, β ring $\underline{H}$) and 7.14 ppm (d,1,J=1.5 Hz, α ring H); mass spectrum (70 ev) m/e (rel. intensity) 228 (6), 227 (10) 226 (61), 115 (4), 114 (16), 113 (100), 85 (13).

bis (2,5-Dimethyl-3-furyl) sulfide (10c) and bis (2,5-dimethyl-3-furyl) disulfide (10d): A mixture of 175 g of 2,5-dimethyl-furan (9) and 0.4 g of stannic chloride was cooled to -20° and 75.3 g of sulfur

dichloride was added while maintaining a temperature
of -20°. After stirring at -20° for 100 mins., the
reaction mixture was warmed to 34° and poured into
ice water (1 liter). Extraction with hexane, and concentration in vacuo, after drying over sodium sulfate,
gave 64.8 g of residue. Column chromatography of the
residue on 1625 g of silicic acid with ether/hexane
(1:20) gave 14.4 g of a mixture of 10c and 10d. Distillation of 11 g of the mixture gave 3.8 g of bis
(2,5-dimethyl-3-furyl) sulfide, b.p. 81-85°C at 0.15
torr, and 5.1 g of bis (2,5-dimethyl-3-furyl) disulfide, b.p. 112-116°C at 0.45 torr.

   10c: ir (neat, KBr plates) 3100, 2940, 2910,
1600, 1560, 1425, 1372, 1330, 1220, 1110, 1060, 979,
918, 790 cm^{-1}; nmr (CCl$_4$) δ 2.2 (s,3, α ring CH$_3$), 2.3
(s,3, α ring CH$_3$) and 5.78 ppm (s,1, β ring H); mass
spectrum (70 ev) m/e (decreasing intensity, mol. ion)
43, 222, 126, 125, 179, 96, intensity of P+2 is 5% of
P.

   10d: ir (neat, KBr plates) 3100, 2942, 2910,
1598, 1560, 1423, 1372, 1330, 1218, 1110, 1060, 979,
918, 790 cm^{-1}; nmr (CCl$_4$) δ 2.1 (s,3, α ring CH$_3$), 2.27
(s,3, α ring CH$_3$) and 6.0 ppm (s,1, β ring H); mass
spectrum (70 ev) m/e (decreasing intensity, mol. ion)
127, 43, 254, 128, 39, intensity of P+2 is 8% of P.

   Repeating the above experiment but replacing the
sulfur dichloride with 98.5 g of sulfur chloride gave
a 54 g residue which when chromatographed gave 13.9 g
of mono- and disulfide. Distillation of 11 g of the
mixture gave 1.96 g of a 40/60 mixture of mono- and
disulfide and 6.6 g of bis (2,5-dimethyl-3-furyl) disulfide boiling at 115° at 0.45 torr.

   bis (2-Methyl-3-furyl) trisulfide (10a): To a
mixture of 5 g of 2-methyl-3-furanthiol (1), 8.7 g of
sodium bicarbonate and 25 ml of ether cooled to -30°
was added a solution of 2.24 g of sulfur dichloride in
20 ml of ether during 24 min. After standing 5 min
the mixture was allowed to warm to -5° and poured into
125 ml of ice water. Extraction with ether (2 x 25 ml),
drying the ether extracts (sodium sulfate) and concentration in vacuo gave 5.1 g of crude 10a. Chromatography on 15 g of silicic acid with hexane gave 3.6 g
of bis (2-methyl-3-furyl) trisulfide as a light yellow
oil, homogeneous by tlc and having: ir (neat, KBr
plates) 3120, 2950, 2919, 1580, 1512, 1384, 1225,
1122, 1085, 935, 885, 730 cm^{-1}; nmr (CDCl$_3$) δ 2.36
(s,3, α ring CH$_3$), 6.35 (d,1,J=2H$_2$, β ring H) and 7.21
ppm (d,1,J=2Hz, α ring H); mass spectrum (70 ev) m/e
(decreasing intensity, mol. ion) 113, 43, 258, 45, 51,
145, 226, 194.

bis (2-Methyl-3-furyl) tetrasulfide (10b):   To a
mixture of 1.65 g 2-methyl-3-furanthiol (1), 10 ml
ether and 3 g sodium bicarbonate cooled to -30°C was
added dropwise a solution of 1.0 g of sulfur monochloride in 10 ml ether. After standing 45 min at
-30° the reaction mixture was poured into water
(75 ml), the upper layer was separated and washed
with 25 ml water. After extracting the combined
aqueous and wash layers with 25 ml ether, the ether
layers were combined, washed with water (2 x 25 ml)
and dried (sodium sulfate). Solvent removal in vacuo
gave 1.6 g of crude 10b. Chromatography on 60 g of
silicic acid with hexane gave 1.1 g of bis (2-methyl-
3-furyl) tetrasulfide as a light yellow oil homogeneous by tlc and having: ir (neat, KBr plates) 3100,
2900, 1570, 1510, 1435, 1380, 1225, 1122, 1086, 938,
887, 730 cm^{-1}; nmr (CCl$_4$) δ 2.37 (s,3, α ring CH$_3$) 6.38
(d,1,J=2Hz, ß ring H), 7.20 ppm (d,1,J=2Hz, α ring H);
mass spectrum (70 ev) m/e (decreasing intensity, mol.
ion) 113, 114, 43, 45, 51, 226, 85, 69, 64, 290, 145.

5-Methyl-2-furanthiol (12): Preparation of 12
according to Goldfarb and Danyushevskii (9) gave,
from 45.1 g of 2-methylfuran (11), 5.9 g of 5-methyl-
2-furanthiol boiling at 56.5° at 33 torr (lit. 44-46°
at 17 torr, (9)) and having: ir (neat, KBr plates)
3130, 2965, 2940, 2538, 1590, 1510, 1450, 1345, 1222,
1200, 1120, 1024, 960, 930, 790 cm^{-1}; nmr (CCl$_4$) δ 2.4
(s,3, α ring CH$_3$), 3.4 (s,broad,1, α ring SH), 6.15
(m,1, ß ring H) and 6.58 (m,1, ß ring H); mass spectrum
(70 ev) m/e (decreasing intensity, mol. ion) 114, 43,
71, 53, 85, 86, P+2 intensity 5.5% of P.

bis (5-Methyl-2-furyl) disulfide (13): A 3.0 g
sample of 5-methyl-2-furanthiol (12) was oxidized with
an aqueous potassium iodide/iodine solution as in the
preparation of 2. A 2.6 g sample of crude 13 was obtained. Chromatography on 81 g of silicic acid with
ether/hexane (1:9) as eluent gave 2.2 g of bis (5-
methyl-2-furyl) disulfide (13) as a bright yellow oil
having: ir (neat, KBr plates) 3135, 2965, 2938, 1590,
1485, 1450, 1224, 1198, 1118, 1022, 960, 929, 790 cm^{-1};
nmr (CCl$_4$) δ 2.47 (s,3, α ring CH$_3$), 6.22 (m,1,J=3Hz,
ß ring H), 6.66 (d,1,J=3Hz, ß ring H); mass spectrum
(70 ev) m/e (decreasing intensity, mol. ion) 114, 113,
43, 71, 53, 85, 84, 226, 150, 160, 194, P+2 intensity
8.5% of P.

2,5-Dimethyl-3-furanthiol (8): A mixture of 35 g
of 2,5-dimethyl-3-thioacetylfuran (10) and 350 ml of
15% aqueous sodium hydroxide solution was heated at reflux for 1 hr, cooled to room temperature, acidified
to pH 1 with 20% sulfuric acid, and extracted with
ether (3 x 100 ml). The combined extracts were washed

with sat. salt solution (4 x 75 ml) and dried (sodium sulfate). Solvent removal in vacuo gave 26 g of crude 8 which on distillation gave 17.3 g of 2,5-dimethyl-3-furanthiol boiling at 79° at 43 torr and having: ir (neat, KBr plates) 3110, 2980, 2950, 2920, 2530, 1610, 1572, 1445, 1430, 1378, 1222, 1063, 982, 920, 790 cm^{-1}; nmr (CDCl$_3$) δ 2.2 (s,3, α ring CH$_3$), 2.25 (s,3, α ring CH$_3$), 2.59 (s,1, β ring SH) and 5.86 ppm (s,1, β ring H); mass spectrum (70 ev) m/e (decreasing intensity, mol. ion) 43, 128, 85, 127, 95, 113 (P+2 intensity 4.5% of P).

Literature Cited and Notes

1. Wilson, R.A., Mussinan, C.J., Katz, I., and Sanderson, A. *J. Agr. Food Chem.*, **21** (5) 873 (1973).
2. Mussinan, C.J., and Katz, I., *J. Agr. Food Chem.*, **21** (1) 49 (1973).
3. Persson, T., and von Sydow, E., *J. Food Sci.*, **38**, 379 (1973).
4. Schutte, L., *Crit. Rev. Food Technol.* 458, Mar. 1974.
5. van den Ouweland, G.A.M., and Peer, H. G., *J. Agr. Food Chem.*, **23** (3) 501 (1975).
6. Giacino, C., *U.S. Patent* 3,394,015 (1968).
7. Scully, J.F., and Brown, E.V., *J. Org. Chem.* **19**, 894 (1954).
8. The identification of 2-methyl-4-furanthiol has been reported by van den Ouweland and Peer (5) as one of the products from the reaction of H$_2$S and 4-hydroxy-5-methyl-3 (2H) dihydrofuranone. However, in the analytical data presented by these authors there is no mention of the intensity of the P+2 ion relative to the parent which would confirm the presence of sulfur and also in the infrared there is no listing of absorbency near 2540 cm^{-1} which is typical of a thiol function. Although the authors claim the structure by comparison of their spectra with those of a synthesized reference compound, no details of the synthetic route are given.
9. Goldfarb, Ya. L. and Danyushevskii, Ya.L., *USSR Patent* 157696, Oct. 14, 1963.
10. Evers, W.J., et al, *U.S. Patent* 3,872,111 (1975).

12

Cat Neural Taste Responses to Nitrogen Compounds

JAMES C. BOUDREAU, JOSEPH ORAVEC, and WILLIAM ANDERSON,
Sensory Sciences Center, University of Texas at Houston Graduate School of
Biomedical Sciences, Houston, Tex. 77025

VIRGINIA COLLINGS, Department of Psychology, University of Pittsburgh,
Pittsburgh, Pa. 15260

THOMAS E. NELSON, Department of Rehabilitation, Baylor College of
Medicine, Houston, Tex. 77025

The cat, Felis catus, is a small carnivore of the family
Felidae, a family which includes closely related species such as
the bobcat, lynx, lion and tiger. As a carnivore and a felid
the cat shares certain common characteristics with these wild
species. The primary behavioral distinguishing characteristic
of the carnivores, and especially of the felids, is their ability
to kill and eat other animals.

In the natural nutritional ecosystem of the felids, vegetable material contributes little directly to the nutrition of
the animal (1, 2, 3). Thus the cat's natural diet consists
largely of small animals, with small mammals and birds predom-
inating and reptiles and invertebrates playing a lesser role.
If small enough, the prey is eaten whole; if large, most of the
soft tissues are consumed.

The major elements of a cat's ecosystem can be considered
in terms of a flow diagram in which the cat can be black boxed
and the input and output signals indicated (Figure 1). In-
putting into the cat are a variety of small organisms and out-
putting are various byproducts, such as feces, urine, volatiles
and eventually the cat itself. These inputs and outputs can be
considered complex chemical signals and the species itself as a
transforming element.

In any natural nutritional ecosystem, various animal species
will be available as foodstuffs to the cat. These animal species
will vary widely in their chemical composition, and some may
even be poisonous to the cat. To aid in the selection of an
adequate diet, the cat is equipped with a wide variety of chemical sensory systems at its oral input end. These sensory
systems monitor various aspects of the chemical nature of the
food input signals, with some systems maximally responsive to
volatiles and others to water soluble compounds (4, 5, 6).

The chemical senses of the tongue are maximally responsive
to water soluble compounds found in foods (7). That these
tongue sensory systems are used in the selection of foodstuffs

of animal origin has been demonstrated by Cott (8, 9) who found that there exists wide variability in the taste of the flesh and eggs of birds, and that cats and humans agree on their preferences. Presumably there also exist wide variations in the palatability of mammalian and invertebrate tissues. The kinds of nutrients eaten by the cat would in part be determined by the sensory neural responses to their chemical composition.

The types of chemoreceptors found on the cat's tongue are similar in most respects to those found on the tongues of other mammals. The chemical sensory systems which have been the most studied are those associated with taste buds: receptor complexes distributed on fungiform papillae on the front part of the tongue and on the back of the tongue in the pits of the vallate papillae. Foliate papillae taste systems are apparently absent in the cat (10).

The taste buds of the fungiform papillae are innervated by neurons that have their cell bodies in the geniculate ganglion. By placing microelectrodes into this ganglion we can record neural pulse signals from single neurons (Figure 2). These pulse signals represent neural signals conveying information concerning the chemical nature of the input signal. By recording pulses when different chemicals are placed on the tongue we can utilize changes in pulse trains to define the nature of active chemical parameters. In this manner we can empirically describe our input variables.

By utilizing various physiological measures (11) we have determined that there are at least three distinct neural groups that innervate fungiform papillae of the tongue. This report will concentrate on group I units, preferentially innervating fungiform papillae on the back part of the tongue, and on group II units, preferentially innervating papillae on the front part of the tongue (Figure 3). Neural groups similar to geniculate ganglion groups I and II have been reported in cat chorda tympani fiber studies (12, 13, 14) and in the dog geniculate ganglion (15). We know relatively little about the chemical stimuli of group III units. In the pages to follow the responses of group I and group II units to a wide range of compounds will be examined (in particular, nitrogen compounds which exhibit maximum stimulability for carnivore taste systems).

Both group I and group II units are sensitive to water solutions of animal tissues. When chicken, pork, beef, fish etc. are chopped up and mixed with distilled water, the solutions are often potent stimuli for eliciting neural discharge from geniculate ganglion units when applied to the appropriate tongue receptive field (Figure 4). Measures such as those in Figure 4 indicate that these chemical sensory systems are concerned with foodstuffs, but these stimuli do not define our input signals precisely enough. The water soluble substances extractable from tissues are of wide variety, although the bulk consists of

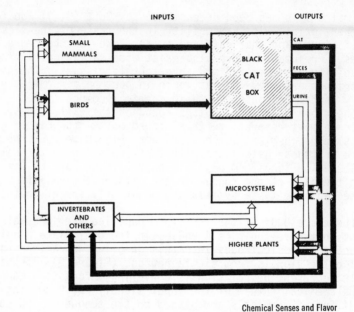

Chemical Senses and Flavor

*Figure 1.   Diagrammatic representation of the flow of nutrients
in the cat's natural nutritional ecosystem (simplified) (4)*

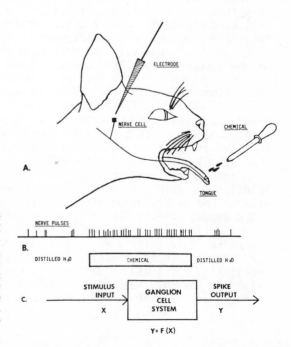

*Figure 2.   A. Diagram of the
experimental setup. Single unit
spike responses are recorded
from neurons in the geniculate
ganglion of the cat while chemical solutions are applied to the
tongue. B. A representation of a
stimulus-evoked single unit spike
train. C. A simplified schematic
of ganglion cell system. The output spike train is functionally
related to the input stimulus.*

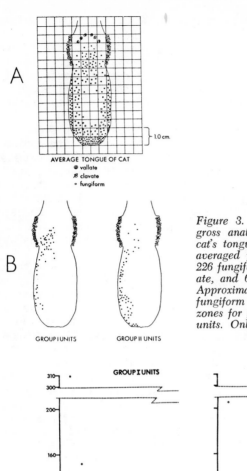

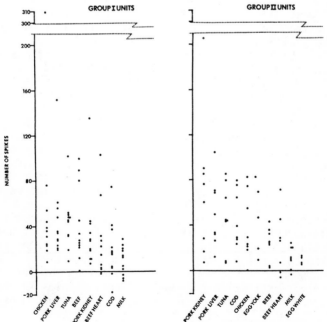

Figure 3. A. Diagram of the gross anatomy of the average cat's tongue. Numerical values averaged from 10 cat tongues: 226 fungiform papillae; 34 clavate, and 6 vallate papillae. B. Approximate locations of the fungiform papillae projection zones for group I and group II units. Only one dot is used for each cell.

Figure 4. Responses of group I and group II units to distilled water solutions of various animal tissues

inorganic salts, organic acids, amino acids, peptides, proteins, nucleotides and nucleotide byproducts. To determine which compounds might be stimulatory, a great many commercially available compounds were tested in 50mM concentration (or less) in distilled water or 50mM saline. In this report, certain stimulus response properties (particularly with respect to nitrogen compounds) will be summarized. More detailed reports are available elsewhere (16, 17).

Group II units innervate chemoreceptive elements on fungiform papillae on the tip and sides of the tongue with medium size fibers. Group II units are the only units that fairly consistently display high rates of spontaneous activity (5 spikes/sec or greater) and therefore can exhibit an increase (excitation) or decrease (inhibition) in spike rate with chemical stimulation. Distilled water by itself is frequently inhibitory. Monovalent salts such as NaCl and KCl are excitatory but noticeable discharge only occurs at levels above 0.1M. Sodium phosphate (NaH_2PO_4) is a potent stimulus at 50mM concentration.

In a study of the responsiveness of group II units to a wide range of chemicals found in animal tissues, it was found that two major classes of compounds affected group II unit discharge: sodium salts of di- and tri-phosphate nucleosides and certain amino acids (Figure 5). Group II units are the only units that respond to a variety of nitrogen compounds. The response to nucleotides seems in part independent of the response to the amino acids, upon which we shall report in some detail.

Amino acids may excite or inhibit group II units. Those compounds that inhibit are L-tryptophan, L-isoleucine, L-tyrosine, and L-phenylalanine. Those compounds that excite are L-proline, L-cysteine, L-lysine, L-ornithine, L-histidine, and L-alanine (4). The neural response to amino acids is stereospecific in that the response may be different to the D- and L- forms. Thus the response to D-proline and D-histidine is but a fraction of that to the L-forms (16).

Upon testing a variety of compounds related to proline and histidine, it was found that the heterocyclic ring components were as active as the parent amino acids. Thus the heterocyclic ring pyrrolidine was as stimulatory as L-proline and the imidazole ring was as active as L-histidine. These results suggested that the activity of these two amino acids resided in the nitrogen heterocyclic rings.

To further explore the stimulability of heterocyclic ring compounds on group II unit discharge, a wide range of heterocyclic compounds were tested. It was found that heterocyclic rings with only heteroatoms of oxygen were inactive. The few sulfur compounds tested were often stimulatory but in a complex manner (e.g., excitation followed by inhibition). The most active compounds discovered were four to six member nonaromatic ring compounds with a single nitrogen. Imidazole was the exception in that it is aromatic with two nitrogens. The most stimulatory heterocycles were 3-pyrroline, pyrrolidine, morpholine, azetidine carboxylic acid, imidazole, piperazine, and piperidine. The five member, single nitrogen, aromatic compound pyrrole was strongly inhibitory. In general, any substituent added onto the ring lowered group II response.

The results of the nitrogen heterocyclic experiment on group II neurons are summarized in Figure 6, a plot of the neural response against the pKa of the nitrogen heterocycles. All compounds were tested at a pH of 7.0 in 50mM saline. Response seemed to be determined by two factors: relative basicity and a steric factor. Compounds with low pKa values were inhibitory or nonstimulatory. As pKa values increase above 5.0 to about 10.0, so also does the neural response. The relative basicity of the compounds is only part of the story however: a steric factor comes into play at pKa values of 11.0. As the ring size increases from 5 to 8, the neural responses changes from excitation to inhibition. Thus pyrrolidine and piperidine are excitatory but azacycloheptane and azacyclooctane are inhibitory.

Group I neurons, those innervating chemoreceptors at the back and sides of tongue, also respond to a variety of chemical solutions. They are the only units discharged by distilled water preceeded by saliva, saline, or Ringer's solution. They are only marginally responsive to NaCl and KCl; although at concentrations of 50mM, NaCl will eliminate the discharge to distilled water. Group I units are responsive to inorganic acids and a wide variety of organic compounds, including many organic acids, certain di- and tri- phosphate nucleosides and a few other compounds (Figure 7).

In a study (17) on a wide variety of compounds mixed in distilled water, the most stimulatory compounds turned out to be compounds with carboxylic and phosphoric acid residues and a small group of nitrogen compounds. The response to all compounds showed a strong pH dependency. Maximum response occurred when the pH of the solution was at or below the pKa of the active group. Thus at a pH of 7.0 malic acid (pKa 3.5, 5.1), a prominent stimulus at lower pH's was entirely ineffective. Similarly, compounds with phosphate groups showed maximum activity when the solution pH was below 5.0.

Although only a few nitrogen compounds were stimulatory for group I neurons, they were highly stimulatory in the pH range from 5.0 to 7.0. The most active compounds proved to be nitrogen heterocycles or to have a nitrogen heterocyclic ring as a component. Many compounds with an imidazole ring proved to be highly effective stimuli. Thus L- and D-histidine, 1-methyl-L-histidine, 1-methyl-imidazole and some histidine dipeptides, including anserine and carnosine, proved to be stimulatory at a

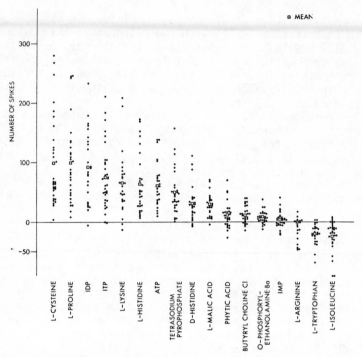

Figure 5. Responses of group II units to a variety of chemicals (the test series) in distilled water in 50mm concentration

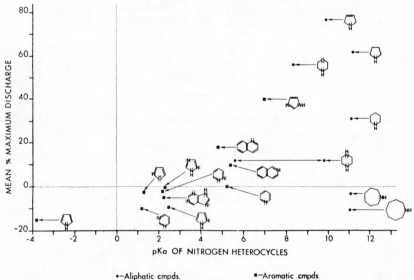

•—Aliphatic cmpds.   ■—Aromatic cmpds.

Figure 6. Relationship between mean percent maximum discharge of group II units and pK$_a$ values of nitrogen and nitrogen–oxygen heterocycles. Values below dotted line indicate inhibition, where inhibition is defined as a decrease in spontaneous activity. Chemicals in 50mm saline at pH 7.0 (16).

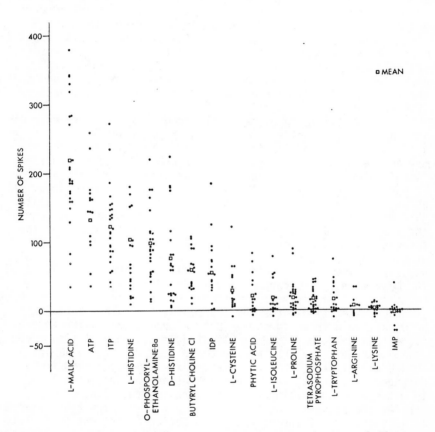

Figure 7.   *Responses of group I units to a variety of chemical compounds (the test series) in distilled water at 50mm concentration*

pH of 7.0. In addition, many of these solutions showed a sharp increase in stimulation when pH was lowered (Figure 8). Imidazole compounds, particularly anserine and carnosine, are of widespread occurrence in animal tissues (18, 19).

The response of group I neurons to the heterocyclic stimulus series was examined. As can be seen in Figure 9, the response to most compounds was minimal. A peak seems to occur near the pKa of 5.0. with the most stimulatory compounds at a pH of 7.0 being pyridine, fused rings with pyridine, and thiazolidine. As was true with other group I stimuli, the response to pyridine and thiazolidine proved highly pH dependent; as pH decreased from 8.0 to 5.0 response increased from near zero to the largest spike discharges yet elicited from group I neurons (Figure 10).

It is possible to classify all group I stimulatory molecules as Brønsted-Lowry acids; i.e., proton donor molecules. Thus the active molecular species in HCl would be the hydronium ion and the active form of carboxylic and phosphoric acids the undissociated molecular form (Figure 11). Relatively few nitrogen groups seem stimulatory but those that do, possess cyclic amine groups which are maximally stimulatory in the protonated state.

On a purely empirical basis we have determined that many of the compounds naturally present in a cat's diet are stimulatory for geniculate ganglion chemoresponsive neural systems. It is probably appropriate that the cat with his animal diet should possess a taste system oriented toward nitrogenous substances. The most stimulatory compounds discovered to date however are certain five and six member nitrogen heterocycles. Thus the most intense taste stimuli for the cat would most likely contain heterocyclic nitrogen groups. Both nitrogen and sulfur heterocyclic compounds have been implicated in cooked meat flavor (20, 21,22 and papers in this symposium). A pKa factor seems present in the cat since group I units are maximally responsive to nitrogen groups with pKa values less than 7.0 and group II units to nitrogen groups with pKa values greater than 7.0.

At the present time it is difficult to compare among species. It is not known if the neural organization of the cat geniculate ganglion is similar to other mammals. Group I and group II units have been definitely identified only in the dog geniculate ganglion (15). The most distinguishing feature in the carnivore is the differential response of group I and II units to nitrogen compounds, particularly amino acids. It is possible that this distinction will be evident in other species.

The species that has been studied with the widest variety of compounds is the human, although no single unit recordings have as yet been taken in this species. The most common taste response to a nitrogen compound is that it is sweet or bitter. Group II excitors often taste sweet to humans, whereas inhibitors taste bitter (16). Cats when given a choice between saline

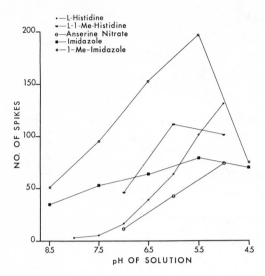

Figure 8. *Discharge of group I units to various compounds with a imidazole group. Solutions varied in pH as indicated. Chemicals in distilled water in 50mm concentration.*

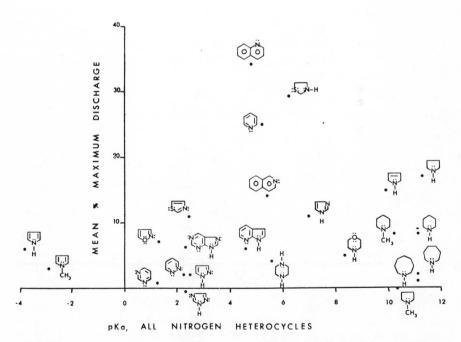

Figure 9. *Relationship between mean percent maximum discharge of group I units and pKₐ values of nitrogen heterocycles. Chemicals in 50mm saline at pH 7.0 (17).*

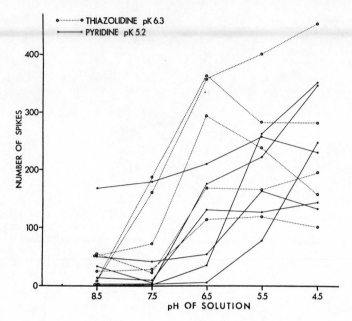

Figure 10.   *Responses of group I units to solutions of pyridine and thiazolidine presented at different pH's* (17)

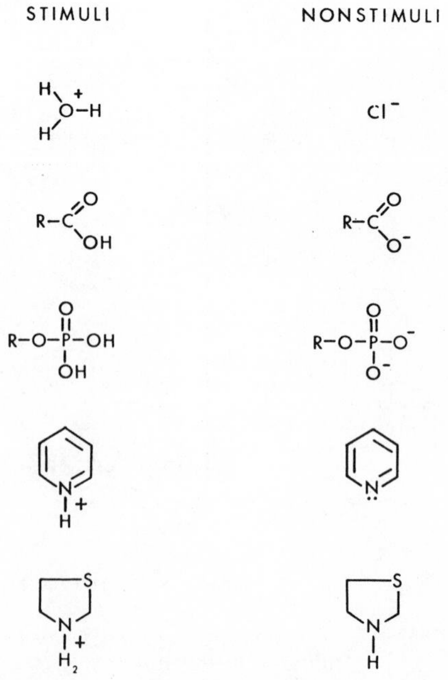

STIMULI   NONSTIMULI

Figure 11. *Chemical groups and their effectiveness in stimulating group I units* (17)

and saline with neurally active compounds preferred the excitors and avoided the inhibitors (23). It is apparent, though, that if a neural system similar to the cat's underlies in part the human sensations of sweet and bitter then there exists great variance in the compounds that stimulate the different species. The compounds that stimulate group I units are in the main quite similar to those eliciting a human sour sensation (17).

Literature Cited

1. Boudreau, J. C., and Tsuchitani, C., "Sensory Neuro-physiology," Van Nostran Reinhold, N.Y., 1973.
2. Ewer, R. F., "The Carnivores," Cornell University Press, Ithaca, 1973.
3. Schaller, G. B., "The Serengeti Lion," The University of Chicago Press, Chicago, 1972.
4. Boudreau, J. C., Chemical Senses and Flavor, (1974) 1: 41-51.
5. Moncrieff, R. W., "The Chemical Senses", Leonard Hill, London, 1967.
6. Tucker, D., In: "Handbook of Sensory Physiology", IV, 151-181, Springer-Verlag, N.Y., 1971.
7. Solms, J., In: "Gustation and Olfaction an International Symposium", 92-110, Academic Press, N.Y., 1971.
8. Cott, H. B., Proc. Zool. Soc. London, (1946) 116:37-524.
9. Cott, H. B., Proc. Zool. Soc. London, (1953) 123:123-141.
10. Sonntag, C. F., Proc. Zool. Soc. London, (1923) 9:291-315.
11. Boudreau, J. C., and Alev, N., Brain Research, (1973) 54:157-175.
12. Cohen, M. J., Hagiwara, S., and Zotterman, Y., Acta Physiol. Scand., (1955) 33:316-332.
13. Nagaki, J., Yamashita, S., and Sato, M., Jap. J. Physiol. (1964) 14:67-89.
14. Ishiko, N., and Sato, Y., Jap. J. Physiol. (1973) 23: 275-290.
15. White, T., and Boudreau, J. C., Neuroscience Abstracts (1975) 1:2.
16. Boudreau, J. C., Anderson, W., and Oravec, J., Chemical Senses and Flavor, in press.
17. Boudreau, J. C., and Nelson, T. E., Manuscript in pre-paration.
18. Crush, K. G., Comp. Biochem. Physiol. (1970) 34:3-30.
19. Suyama, M., Suzuki, T., Maruyama, M., and Saito, K., Bull. Jap. Soc. Sci. Fish (1970) 36:1048-1053.
20. Wilson, R. A., and Katz, I., Flavour Industry (1974) Jan/Feb.:2-8.
21. Gordon, A., Flavour Industry, (1972) Sept.:445-453.

206 PHENOLIC, SULFUR, AND NITROGEN COMPOUNDS IN FOOD FLAVORS

22. Mussinan, C. J., Wilson, R. A., and Katz, I., Agri. and Fd. Chem. (1973) 21:871-871.
23. White, T., and Boudreau, J. C., Physiol. Psychol. (1975) in press.

We thank Coleen Madigan and Jean Watkins for assistance.

This study was financed in part by NSF Grant GMS 73-01296

INDEX

INDEX